High-Temperature Properties and Applications of Polymeric Materials

Martin R. Tant, EDITOR
Eastman Chemical Company

John W. Connell, EDITOR
National Aeronautics and Space Administration

Hugh L. N. McManus, EDITOR
Massachusetts Institute of Technology

Developed from a symposium sponsored
by the Division of Polymeric Materials:
Science and Engineering, Inc.,
at the 207th National Meeting
of the American Chemical Society,
San Diego, California,
March 13–17, 1994

American Chemical Society, Washington, DC 1995

Library of Congress Cataloging-in-Publication Data

High-temperature properties and applications of polymeric materials / Martin R. Tant, editor, John W. Connell, editor, Hugh L. N. McManus, editor.

    p.   cm.—(ACS symposium series; 603)

    "Developed from a symposium sponsored by the Division of Polymeric Materials: Science and Engineering, Inc., at the 207th National Meeting of the American Chemical Society, San Diego, California, March 13–17, 1994."

    Includes bibliographical references and indexes.

    ISBN 0–8412–3313–6

    1. Polymers—Thermal properties. 2. Polymeric composites—Thermal properties.

    I. Tant, Martin R., 1953– . II. Connell, John W., 1959– . III. McManus, Hugh L. N., 1958– . IV. American Chemical Society. Meeting (207th: 1994: San Diego, Calif.) V. American Chemical Society. Division of Polymeric Materials: Science and Engineering. VI. Series.

TA455.P58H58   1995
620.1′9204296—dc20   95–33341
                                                             CIP

This book is printed on acid-free, recycled paper.

PRINTED IN THE UNITED STATES OF AMERICA

1995 Advisory Board

ACS Symposium Series

Foreword

THE ACS SYMPOSIUM SERIES was first published in 1974 to provide a mechanism for publishing symposia quickly in book form. The purpose of this series is to publish comprehensive books developed from symposia, which are usually "snapshots in time" of the current research being done on a topic, plus some review material on the topic. For this reason, it is necessary that the papers be published as quickly as possible.

Before a symposium-based book is put under contract, the proposed table of contents is reviewed for appropriateness to the topic and for comprehensiveness of the collection. Some papers are excluded at this point, and others are added to round out the scope of the volume. In addition, a draft of each paper is peer-reviewed prior to final acceptance or rejection. This anonymous review process is supervised by the organizer(s) of the symposium, who become the editor(s) of the book. The authors then revise their papers according to the recommendations of both the reviewers and the editors, prepare camera-ready copy, and submit the final papers to the editors, who check that all necessary revisions have been made.

As a rule, only original research papers and original review papers are included in the volumes. Verbatim reproductions of previously published papers are not accepted.

Contents

APPLICATIONS AND NEW MATERIALS

Preface

HIGH-TEMPERATURE–HIGH-PERFORMANCE ORGANIC POLYMERS and polymer composites are materials that exhibit superior properties, in terms of both thermal and mechanical behavior, to those of more typical materials. These properties include stability at high temperatures (tens of thousands of hours at 350 °F in air), light weight (low density), high specific strength and stiffness, high toughness, low heat distortion and warpage, and good processability.

The maturation of the field of synthetic polymer chemistry, as well as the development of an improved understanding of structure–property relationships, has resulted in the ability to synthesize materials with properties designed for a particular application. This versatility makes these high-temperature–high-performance organic polymers attractive for aerospace, microelectronic, and other industrial applications. Current uses include films in semiconductor applications, matrix resins in carbon fiber reinforced composites, foams for insulation, ablatives, adhesives for metals and composites, fibers for sporting goods, and membranes for industrial gas separation.

The symposium upon which this volume is based was organized to provide coverage of new polymers as well as the important physics and engineering aspects of these materials relative to processing and performance in various applications. The book begins with an introductory chapter designed to give an overview of the entire field. The remainder of book is organized into three sections: Properties; Processing and Modeling; and Applications and New Materials. Each section contains four to six chapters describing leading edge research and development that encompass a variety of materials, experimental techniques, theories, processes, and applications. Work is included from an international group of scientists and engineers from the United States, Canada, France, Italy, and Australia.

The breadth of the information presented in this book should make it useful for materials scientists, polymer chemists, and engineers in the aerospace, automotive, chemical, and electronics industries. We hope that the book will become a useful resource for a broad spectrum of scientists and engineers whose work concerns the preparation, processing, properties, and applications of polymers and polymer composites.

Acknowledgments

Several organizations generously provided funds to support the symposium upon which this book is based. We thank Eastman Chemical Company, Netzsch, AMETEK—Haveg Division, the Petroleum Research Fund of the American Chemical Society, and the ACS Division of Polymeric Materials: Science and Engineering, Inc.

We very much appreciate the help of Anne Wilson and Rhonda Bitterli of the ACS Books Department who managed to keep both the editors and contributors on track. Finally, we thank the contributors for sharing the results of their work with us.

MARTIN R. TANT
Research Laboratories
Eastman Chemical Company
Kingsport, TN 37662

JOHN W. CONNELL
Langley Research Center
National Aeronautics and Space Administration
Hampton, VA 23681–0001

HUGH L. N. MCMANUS
Department of Aeronautics and Astronautics
Massachusetts Institute of Technology
Cambridge, MA 02139

June 21, 1995

Chapter 1

High-Temperature Properties and Applications of Polymeric Materials

An Overview

Martin R. Tant[1], Hugh L. N. McManus[2], and Martin E. Rogers[1]

[1]Research Laboratories, Eastman Chemical Company, P.O. Box 1972, Kingsport, TN 37662
[2]Technology Laboratory for Advanced Composites, Department of Aeronautics and Astronautics, Massachusetts Institute of Technology, Cambridge, MA 02139

In this chapter we present an overview of the high-temperature properties and applications of polymers and polymer composites. Included are discussions of heat transfer in polymer-based materials, their physical and mechanical properties at elevated temperatures, chemical and physical aging of polymers and effects of aging on properties, and the thermal and mechanical response in high-temperature, high-heat-flux environments. Finally, we present a brief discussion of modern high-temperature polymers and the effect of molecular structure on their properties.

Polymeric materials have long been utilized in high-temperature and high-heat-flux environments. From the natural rubbers first utilized in internal combustion engines earlier in this century to the more modern synthetic polymers used for critical ablative and structural components in recent aerospace applications, polymers have increasingly met the high-performance challenge presented by the design engineer. Much of the credit for this success is due to the maturation of the field of synthetic polymer chemistry. As increasingly more stable polymers have become available, potential engineering applications for these materials have increased dramatically. Development of polymers with improved high-temperature properties has been catalyzed as well by the improved understanding of relationships between polymer structure, both molecular and morphological, and the physical and mechanical properties of these materials. Advances in modeling of high-temperature heat transfer in polymers have aided in the understanding of how polymers react to extreme environments. Finally, advances made in processing of high-performance polymers and their composites continue to trigger improvements in their performance in critical applications.

In this introductory chapter, we present a broad overview of the field of high-temperature properties of polymers and polymer composites. We begin by considering the basic principles of heat transfer in polymer-based materials. The thermophysical properties of these materials relative to their performance in thermal environments are then discussed. The thermomechanical behavior of polymers is then addressed,

0097–6156/95/0603–0001$12.00/0

followed by brief discussions of chemical and physical aging during service and the potential influence of aging on physical and mechanical properties. Having thus established a fundamental understanding of the important aspects of heat transfer and thermophysical properties of polymers, we then describe important aspects of their response in moderately high temperature service and then in extreme environments. Finally, we give a brief overview of the chemistry of modern polymers developed specifically for use in high-performance, high-temperature applications.

Mechanisms of Heat Transfer in Polymers and Polymer Composites

Heat transfer in solid, liquid, and gaseous media occurs by conduction, convection, and radiation. The mechanism of overwhelming importance in non-decomposing polymer-based materials is conduction. Fourier's law of one-dimensional, steady-state heat conduction in an isotropic medium is given as

$$q = -k\frac{dT}{dx} \tag{1}$$

where q is the heat flux in the x direction resulting from the temperature gradient dT/dx, and k is the thermal conductivity. For composites, the thermal conductivity is generally anisotropic, and this must be considered in heat conduction problems. Heat transfer within a thermally decomposing polymer also occurs by convection or flow of decomposition gases within the material. Though radiant heat transfer within a polymer or polymer composite is negligible, radiant heat transfer between the material and its surroundings may indeed be important. This problem is well described in standard heat transfer treatises ($1,2$).

Convective heat transfer between a surface and a flowing fluid (such as a flowing polymer melt) is usually described by

$$q = h(T - T_g) \tag{2}$$

where T is the temperature of the contacting surface, T_s is the temperature of the liquid, and h is the convective heat transfer coefficient. Winter (3) has suggested that the concept of the convective heat transfer coefficient (and thus the Nusselt number) cannot be applied to dissipative flows such as flowing polymer melts. This is illustrated by the fact that for highly dissipative flows the convective heat transfer coefficient may actually turn out to be negative, which, of course, is physically meaningless. Winter uses the Biot number (often used for the thermal boundary condition for solids) for polymer melts.

The energy equation is general and describes energy conservation in any material or process. It is given by

$$\frac{DU}{\rho Dt} = -(\nabla \cdot q) - P(\nabla \cdot v) - (\tau : \nabla v) + S \tag{3}$$

where D/Dt is the substantial derivative, defined by

$$\frac{D}{Dt} = \frac{\delta}{\delta t} + v \cdot \nabla \qquad (4)$$

and ∇ is the differential operator

$$\nabla = \delta_1 \frac{\partial}{\partial x_1} + \delta_2 \frac{\partial}{\partial x_2} + \delta_3 \frac{\partial}{\partial x_3}. \qquad (5)$$

U is the specific internal energy (per unit mass), ρ is the density, t is time, and $\mathbf{q}$ is the heat flux vector. The single term on the left side of equation (3) represents the rate of internal energy gain per unit volume. The term $-(\nabla \cdot \mathbf{q})$ is the rate of internal energy input by conduction per unit volume. The term $-P(\nabla \cdot v)$ is the reversible rate of internal energy increase per unit volume by compression, and the term $-(\tau : \nabla v)$ is the irreversible rate of internal energy increase per unit volume by viscous dissipation. Finally, S is the thermal energy source term that accounts for curing of thermosets, crystallization and melting of thermoplastics, and thermal decomposition. The equation of energy is coupled with the equation of motion through the viscous dissipation term as well as the temperature-dependent viscosity. These two equations must be solved simultaneously along with the equation of continuity. This coupling can make the modeling of polymer processing operations quite complex. But the most complicating factor is that Newton's law of viscosity does not adequately describe the relationship between stress and strain rate in polymers due to the highly viscoelastic nature of these materials. Much more complex constitutive equations must be used (4).

For a fluid at constant pressure and with constant thermal conductivity, and neglecting viscous dissipation, equation (3) reduces to

$$\rho C_p \frac{DT}{Dt} = k \nabla^2 T \qquad (6)$$

where C_p is the specific heat at constant pressure. For a fluid at rest as well as for a solid

$$\rho C_p \frac{\partial T}{\partial t} = k \nabla^2 T \qquad (7)$$

This equation is typically used, along with the energy source term and appropriate initial and boundary conditions, to model polymer heat transfer problems which do not involve fluid flow, e.g. the curing of a thermoset, crystallinity development during molding, and the response of a solid to a high heat flux. More detailed discussions of the equations of change and their applications to polymers and polymer composites may be obtained elsewhere (4-7).

Thermophysical Properties of Polymers and Polymer Composites

The specific heat, C_p, thermal conductivity, k, and thermal diffusivity, α, are thermophysical properties which must be known in order to adequately model the thermal behavior of polymers and composites during processing as well as during application in thermal environments. For processing of thermosets, the heat of reaction must be taken into account, and for thermoplastics, the energy of melting and crystallization is important. For high-temperature applications, the kinetics of decomposition and the thermochemical expansion (resulting from production of pyrolysis gases) must be accounted for in any analysis. The thermal conductivity is defined by Fourier's law for one-dimensional steady-state heat conduction, i.e. equation (1). The thermal diffusivity is defined by

$$\alpha = \frac{k}{\rho C_p} \tag{8}$$

where ρ is the density of the material.

Specific Heat. The specific heat of amorphous polymers is usually in the range of 0.7-2.5 kJ/kg-K, although specific heats as low as 0.2 and as high as 2.7 kJ/kg-K have been observed for some polymers. The specific heat may decrease on crystallization, while that for thermosetting polymers may be affected by the degree of cure. Generally, the specific heat of polymer composites is a weighted average of the components.

—— Differential scanning calorimetry (DSC) is routinely used to measure the specific heat of these materials (8). This technique is also used to study characteristics of the glass transition and to measure the energetics of curing or decomposition as well as the energetics of phase changes such as crystallization and melting. Because of the small sample size (~10 mg), the analyst must take care to ensure that the composition of a polymer composite is representative of the material. While the application of DSC below the onset of thermal decomposition is straightforward, its use above this temperature requires special considerations due to mass loss. Brennan *et al.* (9) developed an iterative method to handle this problem, and this was later modified by Henderson *et al.* (10) to apply actual weight loss data.

Thermal Conductivity and Diffusivity. Heat conduction in polymers has been described by Choy (11). Transfer of heat in solids is typically thought of as occurring by means of molecular vibrations. Theoretical approaches have been mainly concerned with crystalline solids where the lattice vibrations can be resolved into normal modes, the quantization of which leads to the concept of phonons. The problem of calculating thermal conductivity normally reduces to calculation of the number of phonons and their mean free path (12). Unfortunately, this approach is not directly applicable to polymers, since these materials are generally amorphous or only partially crystalline because of their long-chain nature and tacticity which may be unfavorable for crystallization.

The very low thermal conductivity of polymers is one of the primary reasons for the very large effects of heat transfer on the structure and properties of these materials. It is also the main reason that polymers and their composites are used in insulating and thermal protection applications. Typically, the thermal conductivity of amorphous polymers is in the range of 0.1-0.2 W/m-K (or J/s-m-K where J/s = W = Watt). For oriented polymers, it has been experimentally determined that thermal conductivity is higher in the orientation direction - that is, along the polymer chains - and is lower in the transverse direction (*13*). Hansen and Bernier (*14*) observed that the thermal conductivity of highly oriented polyethylene at 51.5°C is higher by a factor of 10 in the orientation direction and lower by a factor of 2 in the normal direction as compared to the unoriented material. The propagation of thermal vibrations along a polymer chain is clearly more effective than interchain propagation between adjacent chains. Thus, the anisotropy in the thermal conductivity of oriented polymers is experimentally found to be linearly related to both draw ratio and Young's modulus in the orientation direction. Crystallinity is also observed to increase thermal conductivity, with values as high as 0.5 W/m-K having been observed (*15*).

The incorporation of reinforcing particles or fibers into polymeric materials may influence the thermal conductivity of the system, depending on the size, shape, orientation, distribution, thermal conductivity, and loading of the reinforcement. The nature of the interface between the matrix and reinforcement may be important as well. More specific information on the thermal conductivity of polymer composites is available elsewhere (*16-19*). Progelhof *et al.* (*16*) reviewed the various models that have been proposed for the thermal conductivity of composites. White and Knuttson (*19*) have demonstrated the use of a thermal conductivity tensor in modeling heat conduction in anisotropic polymer composites.

Equipment for measurement of thermal conductivity and diffusivity of polymers and composites is not as widely available as are differential scanning calorimeters. In general, methods for measuring these properties may be classified as either steady-state, in which case equation (4) is applied, or transient. The line source technique is one of the most widely used transient methods, and several workers have applied it to polymers at low temperatures (<200°C) (*20-22*). Henderson and coworkers (*23*) extended the method to temperatures as high as 900°C to obtain thermal conductivities of polymer composites undergoing thermal decomposition. More recently, Lobo and Cohen (*24*) have applied this technique to measure the thermal conductivity of polymer melts. The laser flash method is another widely used technique, and one of the most reliable for measuring thermal diffusivity (*25*). Thompson has reviewed the various experimental methods for measuring the thermal conductivity and diffusivity of polymers (*26*).

Glass Transition, Crystallization, and Melting. In addition to the thermophysical properties of specific heat, thermal conductivity, and thermal diffusivity, polymers have other properties which are extremely important in characterizing their overall thermal behavior. Such properties include the glass transition temperature, T_g, the crystallization and melting temperatures, T_c and T_m, respectively, and the enthalpies of crystallization and melting, ΔH_c and ΔH_m. The glass transition is a kinetic rather than a thermodynamic transition (*27*). At this temperature, there is an increase in enthalpy and specific volume and the polymer molecules become more mobile due to the increased molecular free volume available for motion as well as their higher kinetic energy. (Hill considers the effects of molecular free volume on physical and mechanical properties in Chapter 5.) The glass transition is observed by DSC as almost a step change in the specific heat. As a kinetic transition, the temperature at which the glass transition is observed depends upon kinetic factors such as heating rate. The long-chain nature and polydispersity of polymers complicates the thermodynamic transitions of crystallization and melting. For example, polymers typically display a rather wide range of crystallization and melting temperatures. In addition, the temperature at which crystallization or melting begins may vary greatly depending on thermal history. Certainly the complexity of the thermal behavior of polymers makes the development of accurate thermal models a more challenging task.

Thermomechanical Properties of Polymers and Composites

The thermomechanical properties of polymers have been studied extensively up to the onset of thermal decomposition. Although much work has been done to characterize the thermophysical properties of polymers and composites during thermal decomposition, very little work has been done to characterize mechanical behavior in this temperature region. (Rapoport and Efros address the fracture of polymers due to high-temperature degradation in Chapter 10). As mentioned above, thermal transition temperatures below the onset of decomposition, i.e. the glass transition temperature, T_g, and the melting temperature, T_m, strongly affect the mechanical properties of polymers and their composites. The morphology of the matrix and the type of reinforcement are also determining factors. For a highly crosslinked polymer, the modulus may decrease only slightly at T_g before becoming strongly affected as decomposition begins. On the other hand, an uncrosslinked, amorphous polymer may display a modulus drop of several decades at the T_g. Crystallization may have an effect similar to crosslinking on modulus above the T_g, i.e. the decrease in modulus will be limited by the presence of crystallites. Once these crystals are melted, flow can occur and the large drop in modulus is observed.

The time- and temperature-dependent mechanical behavior of polymers may be studied in a number of ways, including stress relaxation, creep, and dynamic mechanical spectroscopy (27-30). This latter technique has become more widely applied in recent years due to the fact that it easily separates the mechanical response into viscous and elastic components and the equipment necessary for such analysis has become more readily available and easy to use. In Chapter 4, Czarnek considers the measurement of residual strain using moiré interferometry.

An important industrial "property" of polymers is the heat deflection temperature. While not considered a fundamental polymer property, it is widely used to characterize the temperature at which a polymer ceases to display the desired stiffness. The heat deflection temperature is thus useful for comparing the performance of various polymers at elevated temperatures. In this test (31), a rectangular polymer sample is stressed in a three-point bend configuration to give maximum tensile stresses of 455 kPa or 1820 kPa. The material is heated in a heat-transfer medium at 2°C/min and the temperature is recorded at which the sample deflects 0.25 mm. This is, therefore, a type of creep test which includes temperature as a variable.

Aging of Polymers and Polymer Composites

Phenomena which result in changes in polymer properties during application are often referred to simply as "aging." In reality, there are several chemical and physical phenomena which result in polymer property changes. The term *chemical aging* suggests changes in molecular weight or extent of crosslinking, i.e. changes resulting from the breaking or forming of chemical bonds. The term *physical aging* is usually reserved for structural and property changes resulting from the nonequilibrium nature of polymeric glasses. No change in chemical structure occurs during physical aging. In this section we briefly consider examples of the chemical and physical aging of polymers and polymer composites which are important in high-temperature applications.

Chemical Aging. Chemical aging typically refers to property changes resulting from chemical reactions induced by temperature variations, radiation exposure, humidity, weathering, etc. In high-temperature applications we are obviously concerned with changes in molecular weight or crosslink density which result from exposure to elevated temperatures. At a specified temperature, chemical changes may also be affected by other factors such as moisture, radiation, applied stress, etc. One of the best ways to visualize the relationships between time, temperature, and chemical structure for thermosetting systems is the time-temperature-transformation (TTT) cure diagram developed by Gillham and coworkers (*32-34*). A schematic TTT diagram for a typical thermosetting system is shown in Figure 1. In this figure, T_{go} is the glass transition temperature of the uncured reactants, $_{gel}T_g$ is the temperature at which gelation and vitrification coincide, and $T_{g\ infinity}$ is the maximum glass transition temperature of the system, i.e. the T_g at full cure. Distinct regions of matter in this diagram include liquid, sol/gel rubber, gel rubber, sol/gel glass, gel glass, sol glass, and char. The full cure line ($T_g = T_{g\ infinity}$) separates the sol/gel glass from the gel glass and the sol/gel rubber from the gel rubber. Devitrification and char formation are both degradation processes. The solid lines in the liquid region are isoviscous contours. Such a diagram can also be constructed for linearly polymerizing polymers. Any of the phase changes in this diagram which result from chemical reaction may be termed "chemical aging." Thus, chemical aging is a relatively non-specific term.

Physical Aging. When a polymeric material, whether crosslinked or not, is cooled through the glass transition temperature to an arbitrary aging or annealing temperature, T_a, below T_g, the system is not at thermodynamic equilibrium. Although molecular mobility is greatly reduced in the glassy state compared to the rubbery state, it remains finite. The molecules continue to approach the conformation and specific volume corresponding to equilibrium liquid-like packing. This physical aging process results in decrease in specific volume, and hence molecular free volume, and a corresponding decrease in molecular mobility. This decrease in molecular mobility results in changes in mechanical properties such as increasing modulus and yield strength and decreasing toughness (*27,35-39*). In addition to a decrease in specific volume, there is a corresponding decrease in enthalpy, which is observed as an enthalpy recovery peak during a typical DSC experiment. The embrittlement of glassy polymers as a result of physical aging may prove to be a serious engineering problem, the extent of which depends upon the particular application. The rate of physical aging is usually faster the closer T_a is to T_g. Thus, for polymers having different glass transition temperatures and aged at the same T_a, materials having a higher T_g would be expected to age more slowly. For example, while polycarbonate ($T_g = 150°C$) ages relatively slowly at room temperature, poly(ethylene terephthalate) ($T_g = 65°C$) ages more rapidly on a relative

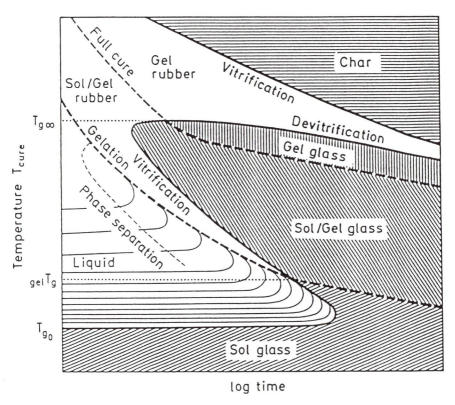

Figure 1.   Schematic time-temperature-transformation (TTT) isothermal cure diagram for a thermosetting system. Abbreviations are defined in the text. (Reproduced with permission from ref. 34. Copyright 1986 Springer-Verlag.)

basis. This might be expected to be one advantage of modern high-temperature polymers, i.e. the higher T_g would result in less physical aging, and thus any resulting property changes, would occur more slowly.

Physical aging of the matrix portion of a polymer composite may also result in changes in mechanical properties of the overall composite (*36-38*). Obviously, the matrix-dominated properties are affected to the greatest extent. The physical aging process is considered in greater depth by Hill in Chapter 5.

Polymer Composites in Moderately High Temperature Service

Polymer matrix composite materials are increasingly used for moderately high temperature (200-350°C) applications. These are applications that do not have extreme temperature requirements, but still involve service temperatures well above the capabilities of typical epoxy matrices in current aerospace use. Advanced polyimide matrices have been formulated for service at these temperatures, and composites using these matrices are already in use in engine cowls, ducts, thrust reversers, and other parts. These composites have been identified as an enabling technology for advancements in turbine engines, hot aircraft structures, and high speed aircraft structures. However, a critical problem preventing more widespread use of these materials is the aging and degradation of the matrix. Continuous exposure to high temperatures is observed to cause mass loss, degradation of properties, shrinkage, and cracking. Lack of capability to quantify matrix degradation and its effect on composite properties and behavior are limiting factors in the application of composite materials to these problems.

Extensive testing has been done to characterize the performance of polymer matrix composites at high temperatures. Recent development efforts in the PMR material system are discussed in (*40*). Most testing has concentrated on isothermal aging, i.e. the effects of continuous soaking at a single high temperature (*41-51*).

It has been observed that high temperature exposure of PMR matrix materials for moderate lengths of time (hundreds of hours) first leads to increases in the glass transition temperature. It is thought that this change is due to additional chemical crosslinking in the matrix. In the presence of air, oxidation may speed this process (*42*). Other changes noted include increases in stiffness, bending strength, and shear strength measured at the exposure temperature. Measured at room temperature, these properties, as well as interlaminar fracture toughness, are reduced (*47,48*). The viscoelastic behavior is also affected, with compliances decreasing with exposure time (*51*).

At the upper end of the temperature range or at longer exposure times, severe mass loss is also noted. Although fibers can lose mass, composite mass loss behavior is dominated by matrix mass loss (*46*). The mass loss is primarily due to oxidative degradation of the polymer, resulting ultimately in its destruction. In non-reactive atmospheres, mass loss is reduced, although not entirely eliminated (*42*). Mass loss is associated with notable shrinkage and loss of mechanical properties. The degradation begins at exposed surfaces, where a distinct degraded layer is observed in both neat resin specimens and composites (*45*). As aging continues, the degraded layer grows. It is suspected that this process is controlled by the rate of diffusion of oxygen into the material (*50*). Eventually, voids and cracks form in the degraded region.

Cracks allow oxygen to penetrate deeper into the material, accelerating the degradation process (45). It is suspected that in extreme cases, an interconnected network of micro-cracks eventually forms, allowing extensive oxidation throughout the laminate, and complete loss of effectiveness of the composite (52).

Attempts have been made to analyze and model various aspects of this problem. Mass loss rates have been empirically fitted to Arrhenius rate curves (49). More sophisticated models have combined modeling of the diffusion of oxygen into the material with chemical reaction rate equations to predict mass loss and the growth of degraded layers (50). Micromechanics and laminated plate theory have been used to predict the effects of matrix shrinkage and property changes on both the degrading layers and the laminate as a whole (53). Cracking of these degraded layers has been predicted by modeling the degraded layers as layers with different material properties in a finite element model (48). At high temperatures, viscoelastic effects are significant. Chemical and physical aging alter the viscoelastic behavior of polymers, and aging-induced changes in viscoelastic properties have been measured and incorporated into standard viscoelastic analysis techniques (51).

Response of Polymer Composites to Extreme Environments

Polymer composites are often exposed to extreme temperature environments, i.e. temperatures higher than the pyrolysis temperature of the polymer. This exposure can be accidental, as in the case of aircraft, civil, or marine structures exposed to fire, or it can be deliberate, as in the case of a charring ablator used as insulation. Such ablators comprise an important aerospace technology, and are used in applications such as rocket motor nozzle linings, reentry heat shields, and blast shields for missile launchers.

Understanding these systems requires simultaneous consideration of energy and mass balances in the pyrolyzing material, as well as models for internal gas release, flow, and pressurization. As mentioned earlier, polymer composites may exhibit anisotropic thermal conductivity, so the conductivity equation, equation (1), becomes

$$q_i = -k_{ij}\frac{\partial T}{\partial x_j} \qquad (9)$$

where q_i is the heat flux in direction i, k_{ij} is the thermal conductivity tensor, and index notation is used, i.e.

$$k_{ij}\frac{\partial T}{\partial x_j} = k_{i1}\frac{\partial T}{\partial x_1} + k_{i2}\frac{\partial T}{\partial x_2} + k_{i3}\frac{\partial T}{\partial x_3} \qquad (10)$$

Pyrolysis generally occurs in-depth, so pyrolysis byproducts, primarily gases, of which water is an important component (54), also exist within the composite. The composite, usually modeled as a homogeneous solid, is assumed to be stationary, but the gases move through the solid. It is assumed that the gases are moving through microscopic pores, and so are in very good contact with the solid. This allows the assumption that the gas and solid temperatures are the same at any given location, although this assumption has been challenged in some work (55). Strains and deformations are assumed small, so the energy balance equation, equation (3), becomes

$$\rho C_p = \frac{\partial}{\partial x_i}\left(k_{ij}\frac{\partial T}{\partial x_j}\right) - C_{p(gas)}m_i\frac{\partial T}{\partial x_i} + RQ \tag{11}$$

The first term on the right is the conduction term. The second is the somewhat confusingly named convection term, in which m_i represents the mass flux of gas in direction i. This term accounts for energy carried through the solid by the moving gas; it is not the convective boundary condition. The final term represents the heat of the pyrolysis reaction, which may be endothermic or exothermic. It is expressed as the product of a reaction rate R and an effective heat of reaction Q.

To solve the above equation, the reaction rates and mass flow rates of gas must be determined. The mass flow rate is determined by a mass balance

$$\phi\frac{\partial \rho_{gas}}{\partial t} = -\frac{\partial m_i}{\partial x_i} + Rn \tag{12}$$

The left hand term represents gas storage, where ϕ is the porosity of the solid. The first term on the right hand side is the divergence of the gas mass flux, and the last term represents gas generation as a constant, n, multiplied by the reaction rate, R. The mass flux is motivated by the pressure created when gas is generated inside the material. The assumed flow of the gas through microscopic pores, cracks or other channels in the material can be described by Darcy's law

$$m_i = -\rho_{gas}\frac{\gamma_{ij}}{\mu}\frac{\partial P}{\partial x_j} \tag{13}$$

where γ is the permeability tensor for the solid material, μ is the gas viscosity, and P is the gas pressure. The gas pressure and density are assumed to be related by the ideal gas law. The reaction rate R is calculated as a function of temperature and pressure using a semi-empirical reaction rate law such as an Arrhenius law. Additional complications are introduced by the fact that the material is undergoing rapid temperature and chemical changes, and the material properties are all functions of the temperature, chemical structure, internal pressure, and stress state. The permeability, for example, changes many orders of magnitude, being very small in typical as-manufactured material, and extremely large in highly charred material (*54*); it is also very sensitive to the stress state, as compressive stresses can close pores and cracks and hence drastically reduce permeability.

Solution of equations (11-13), along with the ideal gas and reaction laws, represents a complex set of coupled non-linear equations. Under some situations, simplifications are possible. If the convection and generation terms are small in equation (11), classic heat conduction solutions can be used. If the composite is undergoing very rapid pyrolysis, internal gas storage can be ignored; if it is degrading slowly under circumstances in which it is tightly constrained, gas flow can be ignored. Study of these two cases have resulted in a very limited set of exact solutions for the pressure distributions (*57*). Generally, numerical methods are required. Solutions have been developed to varying degrees of sophistication by a number of researchers (*56,58-61*).

To predict the performance of the material, it is generally necessary to do a thermostructural calculation as well. The details of such calculations will not be addressed here. The works cited above all include at least some consideration of stresses, including those caused by the internal pressurization due to generation of pyrolysis gases. Considerable work continues in this area (62-65), all based, interestingly, on work originally done on problems in oil and water movement through rock (66).

In practice, mixtures of internal gases are generated by both pyrolysis reaction and desorption of moisture or curing volatiles. The desorption gases are particularly dangerous because they are generated in material that has not yet begun to pyrolyze, and hence is relatively impermeable. The result is extremely high internal pressure, which causes spallation in monolithic materials, delamination or blistering in 2-D composites, and "ply-lift" or "pocketing" failure modes in ablative structures. Depending on the application, this damage may not be fatal to the structure; indeed, blistering may actually help structural survivability temporarily by increasing the insulation value of the material (67). Hence there is a strong motivation to fully understand the behavior of polymeric composites in extreme conditions. This understanding is currently limited by both computational capability and a few areas of insufficient physical knowledge (68). Among these is the exact mechanisms by which moisture or other volatiles are released during the pyrolysis process. Advanced work in this area is the subject of Chapter 7 by McManus and Tai.

Polymers for High-Temperature Applications

There are a variety of polymers available for use as matrix resins in advanced composites. These high-temperature polymers are generally composed of aromatic and heterocyclic moities linked by flexible bridging groups. The polymers generally have high glass transition temperatures, good mechanical properties, and environmental resistance.

Thermosets.   The vast majority of polymers used in advanced composites are thermosetting resins (69). Thermosetting resins generally have low viscosity which facilitates processing and fiber impregnation. However, thermosetting resins must be cured which results in long processing times and makes repair to the final product difficult (70).

Epoxies.   The most widely used matrix resins in advanced composite materials are epoxies (71). The properties of epoxies can be tailored to many applications by using a wide variety of epoxides, curing agents, and additives. In general, epoxies have excellent processing characteristics, good shelf life, and cure without the evolution of volatiles. The thermal properties of epoxy resins vary depending on the chemical composition. For higher temperature applications, the epoxy resins generally contain aromatic or heterocyclic moities. For example, Ciba Geigy's Epoxy MY720™ with the diaminodiphenylsulfone curing agent has a T_g of 180°C and a high modulus and stiffness up to 163°C (71).

EPOXY RESINS
MY 270

Bismaleimides. Bismaleimides (BMI's) are a leading class of thermosetting polyimides. This class of polymers exhibits a balance of thermal and mechanical properties which make them popular for use in advanced composites, structural adhesives, and electronic applications (*72,73*). Bismaleimide resins are of interest because they process in a manner similar to epoxies, have low cure temperatures, and cure without the evolution of by-products (*73,74*). The imide linking groups impart higher polymer backbone stiffness than epoxies, and the greater crosslinking density of BMI's results in materials with improved thermal and hydrolytic resistance relative to epoxies. They have mechanical properties that are similar to epoxies but the thermal stability results in improved performance at high temperatures. However, brittleness and micro-crack formation in cured products are some of the problems attributed to the high crosslink density of bismaleimides (*74*).

BISMALEIMIDES
4,4'-BISMALIMIDODIPHENYL METHANE

PMR Polyimides. PMR (Polymerization from Monomeric Reactants) polyimides are leading thermosetting resins for applications requiring long-term stability at temperatures up to 300°C (*75*). Aromatic imides linked by flexible bridging groups compose the polymer backbone of PMR resins. The aromatic imides impart high glass transition temperatures and thermal stability. The PMR resins are crosslinked through the nadimide endgroups. PMR-15™ composites exhibit greater thermal stability than either bismaleimides or epoxies (*75*). PMR-15™ has a T_g of 345°C (*76*) and retains mechanical properties for hundreds of hours up to 300°C (*75*). PMR-15™ does present processing difficulties such as a cure temperature of 300 to 330°C and batch-to-batch variability. In addition, one of the starting materials, methylene diamine, is toxic. Other PMR resins have been developed with improved thermo-oxidative stability and less toxicity (*76*).

PMR-15™ POLYIMIDE RESIN

Cyanate Esters.   Another class of thermosetting resins for advanced composites is the cyanate esters (77,78). Cyanate ester resins are crosslinked through thermally stable cyanurate rings. These polymers offer several advantages over other thermosetting resins. The neat resins have relatively good toughness, high glass transition temperatures, and low dielectric constants. They absorb smaller amounts of water than other thermosets and demonstrate better long-term hot-wet stability when modified with epoxies (77).

CYANATE ESTERS
2,2'-BIS(4-CYANATOPHENYL)PROPANE

CYANURATE RING

Thermoplastics. High-temperature thermoplastic polymers are composed of aromatic and heterocyclic rings which contribute to the chain stiffness and thermal stability of the polymers. The aromatic and heterocyclic rings are linked by flexible bridging groups which promote processability. In contrast to thermosets, thermoplastics are not crosslinked and promise high volume processing and an associated lower cost per part. High-temperature thermoplastics generally have higher strength, superior impact resistance, and environmental resistance (70). Thermoplastic resins have a longer shelf life and parts made from them may be reheated and reshaped. However, the higher melt temperatures and viscosities make thermoplastic resins more difficult to process in traditional equipment (79).

Polyimides. Among the most thermally stable thermoplastic resins are linear polyimides. Polyimides have excellent mechanical and thermal properties which make them very attractive in a variety of high performance applications. The most common polyimide is PMDA-ODA (Kapton™) polyimide. Kapton™ is a rigid polymer with a high T_g, good thermal stability, and high strength, but it is unsuitable for composite applications due to its insolubility and infusibility. However, polyimides have been developed with high glass transition temperatures, good long-term thermal stability, lower dielectric constants and water absorption, and better processability thus making them suitable as composite matrix resins (*69,80*).

PMDA-ODA POLYIMIDE

Processability of polyimides has been greatly enhanced by incorporating flexible spacers (*81-85*). Hexafluoroisopropylidene, isopropylidene, ether, carbonyl, and sulfonyl bridging groups between rings increase opportunities for bond rotation which decreases the T_g, enhances processability, and generally increases solubility. Also, incorporating non-symmetrical monomers with meta and ortho linkages causes structural disorder in the polymer chain which improves processability (*81,86*).

The amorphous polyetherimides generally have poor solvent resistance, but chemically modified polyetherimides have been developed with improved solvent resistance as well as good processability and good thermal and hydrolytic stability (*75*). The combination of flexible bridging groups and meta linkages is evident in commercial polyetherimides. General Electric's Ultem™ is an amorphous polyimide that can be melt processed at temperatures of 350-425°C (*87*). Ultem™ has a glass transition temperature of 220°C (*75*).

POLYETHERIMIDE

Fluorine-containing polyimides show improved processability, lower dielectric properties, and good optical properties while maintaining excellent thermal stability, Fluorine can be incorporated in bridging groups, bulky sidechains, or as substituents on aromatic rings.   Fluorine-containing polyimides have found application as composite matrix resins (*88*). A fluorine-containing polyimide, Avimid N™, is one of the most thermally stable commercial polyimides available with a maximum use temperature of 316°C (*88*). Fluorinated polyimides have been developed with glass transition temperatures above 400°C and are processable in organic solvents (*89*).

FLUORINE-CONTAINING POLYIMIDE
AVIMID N™

Poly(arylene ether)s.   A prominent class of thermoplastic materials used in advanced composites is the poly(arylene ether)s (*90,91*). The most widely studied poly(arylene ether) is polyetheretherketone or PEEK. This polymer is semicrystalline with a T_g of 145°C and a T_m of 335°C. PEEK is known for its high strength, toughness, and excellent resistance to moisture and solvents.   Other important semicrystalline poly(arylene ether)s include polyetherketone (PEK) which has a T_g of 165°C and a T_m of 365°C and polyetherketoneetherketoneketone (PEKEKK) which has a T_g of 173°C and a T_m of 374°C.

POLYETHERETHERKETONE

Among the amorphous poly(arylene ether)s are the polysulfones including polyethersulfone (PES). Amoco's Udel™ PES has a T_g of 185°C and is generally characterized by its high temperature performance, good mechanical properties, and low flammability, but it has poorer solvent resistance than the semicrystalline poly(arylene ether)s (*91,92*). Amoco's Radel R™ PES does not contain the flexible isopropylidene bridging group. As a consequence, Radel R™ has a higher T_g at 220°C than Udel™ and a higher heat distortion temperature (*92*).

POLYETHERSULFONE
UDEL™

RADEL R

Concluding Comments

Within this brief overview we have presented some of the more important aspects of the high-temperature properties and behavior of polymers and polymer composites, the response of such materials in high-temperature environments, and, finally, a brief look at some of the modern polymers used in high-temperature applications. The chapters which follow will delve more deeply into these areas and, we hope, provide a useful resource for scientists and engineers working in this rapidly evolving field.

Literature Cited

1. Holman, J. P. *Heat Transfer*; McGraw-Hill; New York, NY, 1981.
2. Incropera, F. P.; DeWitt, D. P. *Fundamentals of Heat Transfer*; Wiley; New York, NY, 1981.
3. Winter, H. H. *Adv. Heat Transfer* **1977**, *13*, 205.
4. Bird, R. B.; Armstrong, R. C.; Hassager, O. *Dynamics of Polymeric Liquids*; 2nd Ed.; Wiley; New York, NY, 1987.
5. Bird, R. B.; Stewart, W. E.; Lightfoot, E. N. *Transport Phenomena*; Wiley; New York, NY, 1960.
6. Tadmor, Z; Gogos, C. G. *Principles of Polymer Processing*; SPE Monograph Series; Wiley-Interscience; New York, NY, 1979.
7. Tucker, C. E. III *Fundamentals of Computer Modeling for Polymer Processing*; Hanser Publishers, New York, NY, 1989.
8. Runt, J.; Harrison, I. R. *Meth. Exp. Phys.* **1980**, *16B*, 287.
9. Brennan, W. P.; Miller, B.; Whitwell, J. C. *Ind. Eng. Chem. Fund.* **1969**, *8*, 314.
10. Henderson, J. B.; Wiebelt, J. A.; Tant, M. R.; Moore, G. R. *Thermochim. Acta* **1982**, *57*, 161.
11. Choy, C. L. *Polymer* **1977**, *18*, 984.
12. Klemens, P. G. In *Physics of Non-Crystalline Solids*; J. A. Prins, Ed.; North-Holland/Wiley, Amsterdam/New York, NY, 1965.
13. Choy, C. L.; Luk, W. H.; Chen, F. C. *Polymer* **1978**, *19*, 155.
14. Hansen, D; Bernier, G. A. *Polym. Eng. Sci.* **1972**, *12*, 204.
15. Choy, C. L.; Young, K. *Polymer* **1977**, *18*, 842.
16. Progelhof, R. C.; Throne, J. L.; Ruetsch, R. R. *Polym. Eng. Sci.* **1976**, *16*, 615.
17. Bigg, D. M. *Polym. Eng. Sci.* **1977**, *17*, 842.
18. Bradbury, E. J.; Bigg, D. M. *J. Mechanical Design* **1980**, *102*, 823.

19. White, J. L.; Knuttson, B. A. *Polym. Eng. Rev.* **1982**, *2*, 71.
20. Underwood, W. M.; Taylor, J. R. *Polym. Eng. Sci.* **1978**, *18*, 556.
21. Nix, G. H.; Lowery, G. W.; Vachon, R. I.; Tanger, G.E. Paper No. 67-314, AIAA Thermophysics Spec. Conference, 1967.
22. Greenwood, L. R.; Comparin, R. A. *J. Spacecraft Rockets* **1970**, *7*, 362.
23. Henderson, J. B.; Verma, Y. P.; Tant, M. R.; Moore, G. R. *Polym. Comp.* **1983**, *4*, 219.
24. Lobo, H.; Cohen, C. *Polym. Eng. Sci.* **1990**, *30*, 65.
25. Taylor, R. E. *High Temperatures-High Pressures* **1979**, *11*, 43.
26. Thompson, E. V. "Thermal Properties" In *Encyclopedia of Polymer Science and Engineering*, Vol. 16, Wiley, New York, NY, 1979.
27. Tant, M. R.; Wilkes, G. L. *Polym. Eng. Sci.* **1981**, *21*, 874.
28. Aklonis, J. J.; MacKnight, W. J. *Introduction to Polymer Viscoelasticity*; Wiley, New York, 1983.
29. Ward, I. M. *Mechanical Properties of Solid Polymers*; Wiley, New York, 1993.
30. Ferry, J. D. *Viscoelastic Properties of Polymers*; Wiley, New York, 1980.
31. ASTM D 648-82, *Standard Test Method for Deflection Temperature of Plastics Under Flexural Load.*
32. Gillham, J. K. In *Developments in Polymer Characterization*; Dawkins, J. V., Ed.; Applied Science, London, 1982; Vol. 3.
33. Enns, J. B.; Gillham, J. K. *ACS Adv. Chem. Ser.* **1983**, *203*, 27.
34. Aronhime, M. T.; Gillham, J. K. *Adv. Polym. Sci.* **1986**, *78*, 83.
35. Struik, L. C. E. *Physical Aging in Amorphous Polymers and Other Materials*; Elsevier, New York, NY, 1978.
36. Kong, E. S-W. *Adv. Polym. Sci.* **1986**, *80*, 125.
37. Sullivan, J. L. *Comp. Sci. Tech.* **1990**, *39*, 207.
38. Ogale, A. A. *Comp. Mater. Sci.* **1991**, *7*, 205.
39. Brennan, A. B.; Feller III, F. *J. Rheol.* **1995**, *39*, 453.
40. Meador, M. A.; Cavano, P. J.; Malaric, D. C. In *Proceedings of the Sixth Annual ASM/ESD Advanced Composites Conference*; ASM: Detroit, MI, 1990; pp 529-539.
41. Bowles, K. J. *A Thermally Modified Polymer Matrix Composite Material with Structural Integrity of 371°C*; TM-100922; NASA: Washington, DC, 1988.
42. Vannucci, R. D.; Cifani, D. In *Proceedings of the 20th International SAMPE Conference*; SAMPE: Covina, CA, 1988; pp 562-575.
43. Bowles, K. J.; Nowak, G. *J. Compos. Mater.* **1988**, *22*, 966-985.
44. Bowles, K. J. *SAMPE Quart.* **1990**, *21*, 6-13.
45. Bowles, K. J.; Jayne, D.; Leonhardt, T. A. *SAMPE Quart.* **1993**, *24*, 3-9.
46. Martin, R. H.; Siochi, E. J.; Gates, T. S.; In *Proceedings of the American Society for Composites 7th Technical Conference on Composite Materials*; Technomics: Lanchaster, PA, 1992; pp 207-217.
47. Pederson, C. L.; Gillespie, J. W. Jr.; McCullogh, R. L.; Rothschilds, R. J.; Stanek, S. L. *The Effect of Isothermal Aging on Transverse Crack Development in Carbon Fiber Reinforced Cross-Ply Laminates*; CCM Report 93-42; University of Delaware Center for Composite Materials: Newark, DE, 1993.

48. Pederson, C. L. *The Effect of Temperature on Transverse Cracking in High Performance Composites*, CCM Report 92-28; University of Delaware Center for Composite Materials: Newark, DE, 1992.

49. Grayson, M. A.; Fry, C. G. *On the use of a Kinetic Map to Compare the Thermal Stability of Polymeric Materials Undergoing Weight Loss*, McDonnell Douglas Co. preprint, 1994.

50. Hipp, R. C.; Harmon, D. M.; McClellan, P. S. *Accelerated Aging and Methodology Development for Polymeric Composite Material Systems*, MacDonnell Douglas Co. preprint, 1994.

51. Gates, T. S; Brinson, L. C. In *Proceedings of the 36th AIAA/ASME/ASCE/ AHS/ASC Structures, Structural Dynamics, and Materials Conference*; AIAA: Washington, DC, 1994; pp 2096-2105.

52. Bowles, K. J. *J. Adv. Materials* **1994**, *26*, 23-32.

53. McManus, H. L.; Chamis, C. C. *Stress and Damage in Polymer Matrix Composite Materials Due to Material Degradation at High Temperatures*, in press as a NASA TM, 1994.

54. Sykes, G. F., Jr. *Decomposition Characteristics of a Char-Forming Phenolic Polymer Used for Ablative Composites*; NASA TN D-3810, February 1967.

55. Florio, J.; Henderson, J. B.; and Miriyale, S. K. In *Computational Mechanics of Porous Materials and Their Thermal Decomposition*; Salamon, N. J.; Sullivan, R. M., Eds.; AMD-136; American Society of Mechanical Engineers: New York, 1992, pp 91-101.

56. McManus, H. L.; Springer, G. S. *J. Compos. Mater.* **1992**, *26*, 206-229.

57. McManus, H. L. In *Proceedings of the 1992 JANNAF Rocket Nozzle Technology Subcommittee Meeting*; CPIA Publication 592; Chemical Propulsion Information Agency: Columbia, MD, 1992.

58. Henderson, J. B.; Wiebelt, J. A.; Tant, M. R. *J. Compos. Mater.* **1985**, *19*, 579-595.

59. Henderson, J. B.; Wiecek, T. E. *J. Compos. Mater.* **1987**, *21*, 373-393.

60. Sullivan, R. M.; Salamon, N. J. *Int. J. Eng. Sci.* **1992**, *30*, 431-441.

61. Kuhlmann, T. L. *Thermo-Chemical-Structural Analysis of Carbon-Phenolic Composites with Pore Pressure and Pyrolysis Effects*; Ph.D. Thesis; University of California: Davis, CA, 1992.

62. McManus, H. L. In *Proceedings of the 33rd AIAA/ASME/ASCE/AHS/ASC Structures, Structural Dynamics, and Materials Conference*; AIAA: Washington, DC, 1992; pp 3172-3178.

63. Weiler, F. C. In *Computational Mechanics of Porous Materials and Their Thermal Decomposition*; Salamon, N. J.; Sullivan, R. M., Eds.; AMD-136; American Society of Mechanical Engineers: New York, 1992; pp 1-28.

64. Sullivan, R. M. In *Mechanics of Composite Materials: Nonlinear Effects*; Hyer, M. W., Ed.; AMD-159; American Society of Mechanical Engineers: N. Y., NY, 1993; pp 331-341.

65. Wu, Y.; Katsube, N. In *Computational Mechanics of Porous Materials and Their Thermal Decomposition*; Salamon, N. J.; Sullivan, R. M., Eds.; AMD-136; American Society of Mechanical Engineers: N. Y., NY, 1992; pp 103-112.

66. Biot, M. A. *J. Appl. Phys.* **1955**, *26*, 182-185.

67. McManus, H. L. In *Proceedings of the 9th International Conference on Composite Materials*; Woodhead Publishing: Cambridge, United Kingdom, 1993, Vol. 5; pp 929-936.

68. McManus, H. L. In *Computational Mechanics of Porous Materials and Their Thermal Decomposition*; Salamon, N. J.; Sullivan, R. M., Eds.; AMD-136; American Society of Mechanical Engineers: N. Y., NY, 1992, pp 113-119.

69. Thayer, A. M. *C&E News* (July 23, 1990) 37.

70. Bigg, D. M. In *International Encyclopedia of Composites*; Lee, S. M., Ed.; VCH Publishers: New York, 1990; Vol. 6.

71. Browning, C. E. In *Advanced Thermoset Composites*; Margolis, J. M., Ed.; Van Nostrand Reinhold: New York, 19.

72. Stenzenberger, H. D. In *Polyimides*; Wilson, D., Stenzenberger, H. D., and Hergenrother, P. M., Eds.; Chapman & Hall: New York, 1990.

73. Stenzenberger, H. D. In *Structural Adhesives: Developments in Resins and Primers*; Kinloch, A. J., Ed.; Elsevier Applied Science: New York, 19.

74. Takekoshi, T. *Adv. Polym. Sci.* **1990**, *94*, 1.

75. Wilson, D. In *Polyimides*; Wilson, D., Stenzenberger, H. D., and Hergenrother, P. M., Eds.; Chapman & Hall: New York, 1990.

76. Wilson, D. *High Perform. Polym.* **1993**, *5*, 77-95.

77. Graver, R. B. In *International Encyclopedia of Composites*; Lee, S. M. Ed.; VCH Publishers: New York, 1990; Vol. 1.

78. Srinivasan, S. A.; McGrath, J. E. *High Perform. Polym.* **1993**, *5*, 259-274.

79. Cogswell, F. N. *Thermoplastic Aromatic Polymer Composites*; Butterworth Heinemann: Oxford, United Kingdom, 1992.

80. Economy, J. In *Contemporary Topics in Polymer Science*; Vandenberg, E. J., Ed.; Plenum: New York, 1984; Vol. 5.

81. St. Clair, T. L. In Polyimides; Wilson, D., Stenzenberger, H. D., and Hergenrother, P.M., Eds.; Chapman & Hall; New York, 1990.

82. Mercer, F. W.; Goodman, T. D. *Polym. Prepr.* **1991**, *32*(2), 189.

83. Burks, H. D.; St. Clair, T. L. In Polyimides: Synthesis, Properties, and Applications; Mittal, K. L., Ed.; Plenum: New York

84. Harris, F. W.; Karnavas, A. J.; Das, S.; Curcuras, C. N.; Hergenrother, P. M. *Polym. Mater. Sci. Eng.* **1986**, *54*, 89.

85. Hergenrother, P. M., Havens, S. J. *J. Polym. Sci.: Part A*, **1989**, *27*, 1161.

86. Scroog; C. E. *Prog. Polym. Sci.* **1991**, *16*, 561.

87. Hancox, N. L. In *Concise Encyclopedia of Composite Materials*; Kelly, A., Cohn, R. W., and Bever, M. B., Eds.; Pergamon: Englewood Cliffs, NJ, 1989.

88. Scola, D. A., Vontell, J. *Chemtech*, **1989**, *19*(2), 112.

89. Rogers, M. E.; Brink, M. H.; McGrath, J. E.; Brennan, A. *Polymer* **1993**, *34*, 849.

90. Rosato, Donald V.; DiMattia, D. P.; Rosato, Dominick V. *Designing with Plastics and Composites: A Handbook*; Van Nostrand Reinhold: New York, 1991.

91. Beland, S. *High Performance Thermoplastic Resins and Their Composites*; Noyes Data Corporation: New Jersey, 1991.

92. Rigby, R. B. In *Engineering Thermoplastics*; Margolis, J. M., Ed.; Marcel Dekker: New York, 1985.

RECEIVED June 26, 1995

PROPERTIES

Chapter 2

Thermophysical Properties of Epoxy and Phenolic Composites

R. E. Taylor[1] and R. H. Bogaard[2]

[1]Thermophysical Properties Research Laboratory, School of Mechanical Engineering and [2]CINDAS, Institute for Interdisciplinary Engineering Studies, Purdue University, 2595 Yeager Road, West Lafayette, IN 47906

The thermophysical properties (specific heat, thermal conductivity, thermal diffusivity and thermal expansion) of epoxy and phenolic composites are examined. The most common measurement techniques are briefly discussed. Both original and representative literature data are presented to illustrate observed behavior.

When the thermophysical properties for epoxies and phenolics are compared to those for other materials such as metals and ceramics, it is generally found that the epoxies and phenolics have much lower conductivities and larger specific heats and expansions. It is also noted that the environment and temperature effects on the properties of the epoxies and polymers are much more pronounced than for other material classes. This is due to the molecular structure of these materials and their relatively low melting or degradation temperatures. On a temperature scale normalized to the "melting" temperatures, these materials are used at much higher temperature ratios than other materials. In effect, the situation is comparable to continual use of iron alloys at temperature above 800°C.

The sensitivity of the thermophysical properties of epoxy and phenolics to environmental conditions is shown, for example, by thermal expansion testing. When samples of these materials are placed in a dilatometer and dry gas introduced, the samples often shrink even though the temperature is held constant. This is due to removal of moisture. This phenomena is not usually observed for metals or ceramics.

0097–6156/95/0603–0022$12.00/0

The large difference in the thermophysical properties of polymers compared to those for metal or ceramic reinforcements, combined with the softness of polymers and their easy fabricability, makes it possible to make composites with a wide range of properties.

Thermophysical Properties of Composites: Original Data

Specific Heat. Currently, most specific heat measurements are made using differential scanning calorimetry (DSC), as defined by ASTM E1269-94. Usually sapphire is the reference standard. The standard and sample are subjected to the same heat flux as a blank, and the differential powers required to heat the sample and standard at the same rate are determined using a digital data acquisition system. From the masses of the sapphire standard and sample, the differential power, and the known specific heat of sapphire, the specific heat of the sample is computed. Other test methods, such as adiabatic calorimetry *(1)* or drop calorimetry *(2)*, are much more labor intensive and less convenient *(3)*. Also the accuracy of DSC techniques has increased markedly in recent years and is comparable to all but the most accurate techniques.

The specific heats of polymers are larger than those of other materials due to the low atomic number of the constituents and to molecular bending and rotation. According to the DuLong-Petit Law, the specific heat is 6 calories per mole, so atoms of low atomic weight have large specific heats. While it is not possible to calculate the specific heat of a polymer from first principals (as it is to a certain extent for metals), the Rule of Mixtures is generally valid, so it is possible to calculate the specific heat of a polymer composite from the specific heats of the polymer and the reinforcements. Directionality is not a factor. The Rule of Mixtures for specific heat is expressed as:

$$C = \frac{1}{\rho}(V_f \rho_f C_f + V_m \rho_m C_m) \tag{1}$$

where ρ is the density, V is the volume fraction, C is specific heat and the subscripts f and m denote fiber and matrix properties, respectively. The composite density ρ is itself predicted from the Rule of Mixtures:

$$\rho = V_f \rho_f + V_m \rho_m \tag{2}$$

Figure 1 compares the specific heats of two graphite fiber reinforced PEEK materials. The volume percent of fibers (62%) was the same for both samples, but one was unidirectionally reinforced, while the other was quasi-isotropic. The specific heat values are within a few percent

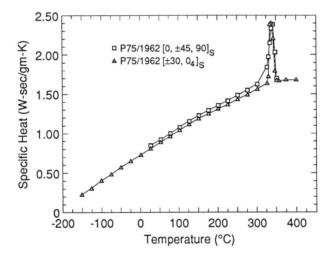

Figure 1. Specific Heat of Graphite/Thermoplastic (62 Vol % P75 in PEEK) with Quasi-Isotropic and Zero CTE Layups.

of those predicted from the Rule of Mixtures. The peaks in the specific heat curve are due to a transition in the PEEK.

It should be emphasized that the Rule of Mixtures applies to true specific heat, i.e. that quantity of heat required to raise one gram of material one degree C. Reported specific heat values are usually apparent specific heats which include energy which changes the material: desorption of moisture, phase changes, energetics of decomposition, etc. In determining specific heat, thoroughly dried samples should be used; otherwise the results depend upon the relative humidity. Also there must be no cutting oil, fingerprints, or other contaminants present. The heat required to remove those contaminants is significant compared to the heat required to raise one gram one degree. Therefore the presence of such contaminants introduces significant errors into the measurements. The small sample mass magnifies the problem.

Thermal Conductivity/Thermal Diffusivity. Thermal conductivity (K) values may be calculated from the results of steady state measurements or from the results of transient measurements. Both methods are indirect. Thermal conductivity is a derived quantity and cannot be measured directly. Transient techniques are often used (4) because of their relative ease, speed and the fact that they can reveal important information concerning the microstructure of the samples *(5).* However, the measurement of the properties of heterogeneous materials by transient techniques requires a great deal of expertise. This is particularly true in the case of fiber-reinforced composites where the fibers act as "super-highways" for heat flow *(6).* Actually, a similar problem exists for steady- state determinations, since the experimenter assumes that the heat flux is uniform or that the surface temperatures are uniform - neither of which is correct. The laser flash diffusivity technique has been established as an ASTM standard (ASTM E1461-9) and its application to all types of heterogeneous materials has been investigated *(6).*

In the flash method, the front face of a small disc-shaped sample is subjected to a submillisecond laser burst, and the resulting rear face temperature rise is recorded and analyzed. A computer controls the experiment, collects the data, calculates the results, corrects for heat losses and compares the raw data with the theoretical model.

Conductivity values for material with continuous fiber reinforcements can be approximated *(7,8)* by the following equations:
   • For longitudinal conductivity,

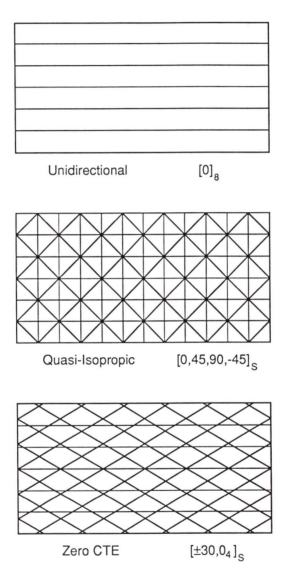

Unidirectional   [0]$_8$

Quasi-Isopropic   [0,45,90,-45]$_S$

Zero CTE   [±30,0$_4$]$_S$

Figure 2. Figure Layup Configurations.

$$K_x = V_f K_{f_x} + K_m (1 - V_f)$$ (3)

• For transverse conductivity,

$$K_y = K_z = \left(1 - \sqrt{V_f}\right) K_m + \frac{K_m \sqrt{V_f}}{1 - \sqrt{V_f} \left(1 - K_m / K_{f_y}\right)}$$ (4)

where K is thermal conductivity and x, y and z refer to direction.

In the case of unidirectional fibers, Eq. (3) is very satisfactory, and Eq. (4) is approximately correct. The conductivity values of multi-layer composites can be calculated from ply values (7).

A thorough study of thermal conductivity, thermal expansion and specific heat of three continuous fiber-reinforced composites using three layups and keeping the fiber volume percent constant at 62% was performed by TPRL (NASA Report, CR-187472, Aug., 1990) (8). The three composite systems were graphite epoxy [P75/ERLX 1962], Pitch-fiber based graphite/thermoplastic [P75/PEEK], and a pan-fiber based carbon/thermoplastic [AS4/PES]. The layups were quasi-isotropic [0,45,90,-45]$_s$ and near-zero CTE [30,-30,0$_4$]$_s$ (Table I and Figure 2). Some measurements were also made on unidirectional composites.

The conductivity values for the P75/PEEK samples are plotted in Figure 3. The Z-direction values are the same and are quite small. The conductivity of the zero CTE layup in the X-direction is quite large, while that for the Y direction is much less, but still considerably larger than those for the Z direction. In the case of the quasi-isotropic layup, the conductivity values in the X- and Y- directions are nearly equal to each other and are midway between the corresponding values for the zero CTE layup.

Thermal conductivity values at 100°C for the three continuous fiber-reinforced composites are summarized in Table I. The Z-direction values are all small, as expected. The conductivity values for both P75 graphite composites are nearly equal in corresponding in-plane directions, demonstrating that the fiber controls the values, and not the matrix. The conductivity values for the in-plane directions in the AS4 composite are much smaller than those for both P75 composites, in line with the previous comment. Both the lack of quasi-isotropic response in [0,45,90,-45]$_s$ and low longitudinal conductivity in [30,-30,0$_4$]$_s$ suggest the presence of disbonds or microcracks in the carbon-reinforced composite (8).

PMCs degrade at high temperatures, permanently altering their conductivities. A knowledge of this degradation is important in applications in which these materials are used as heat shields. In a

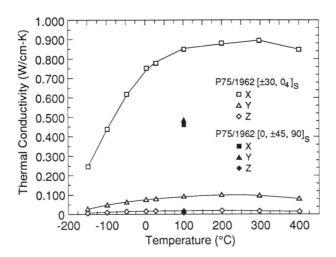

Figure 3. Thermal Conductivity in the X, Y and Z Directions of Graphite/Thermoplastic (62 Vol % P75 in PEEK) with Zero CTE and Quasi-Isotropic Layups.

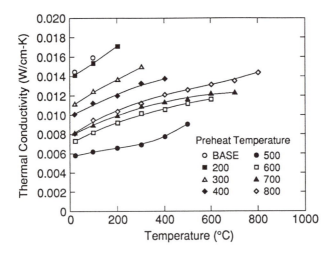

Figure 4. Thermal Conductivity of Carbon-Reinforced Phenolic in the Ply Direction Pre-Heat Treated to Various Temperatures (200, 300, 400, 500, 600, 700 and 800 ° C). Reprinted with permission from Reference 9. Copyright 1989 Plenum Publishing.

proprietary study of a carbon fiber reinforced phenolic composite, conductivity values were determined for samples which had previously been heated for twenty minutes to various temperatures between 200 and 800°C by the manufacturer prior to shipment to the measurement facility (9). The conductivity values for the across-ply direction for many different degradation states are given in Figure 4. By plotting the values measured at the soak temperature (Figure 4), the conductivity of the composite as a function of temperature for twenty minute exposures can be deduced (Figure 5). Alternately, because thin samples are essentially exposed on all surfaces and each diffusivity experiment can be performed rapidly, it is possible to determine changes in conductivity for the composite as it degrades by simply holding the sample temperature constant and measuring diffusivity as a function of time.

Table I. Thermal Conductivity (W/cmK) at 100°C

Material	K_x	K_y	K_z
P75/1962 $[0]_8$	1.05	0.012	0.012
P75/1962 $[0,\pm45,90]_s$	0.50	0.48	0.015
P75/1962 $[\pm30,0_4]_s$	0.97	0.098	0.015
P75/PEEK $[0,\pm45,90]_s$	0.46	0.58	0.015
P75/PEEK $[\pm30,0_4]_s$	0.85	0.082	0.018
AS4/PES $[0,\pm45,90]_s$	0.029	0.050	0.005
AS4/PES $[\pm30,0_4]_s$	0.029	0.007	0.002

Thermal Expansion. Expansion studies can be performed relatively quickly and inexpensively using push-rod dilatometers (ASTM E228). However, the sensitivities of push-rods are not adequate to accurately measure the small changes seen in the small temperature ranges involved. Thus push rod-dilatometers are useful for establishing the overall behaviors of continuous-reinforced polymer composites. Precise measurements should be taken using laser interferometric dilatometry. Contractions can be observed on graphite/polymer samples as the chamber they are in is evacuated or purged with dry gas. These length changes, due to moisture disorption, can be appreciable. In the fiber direction of continuously-reinforced fiber composites, changes can be equal to the dimensional changes caused by the usual temperature cycles investigated. Thus, the sample should be evacuated for several hours before beginning a test.

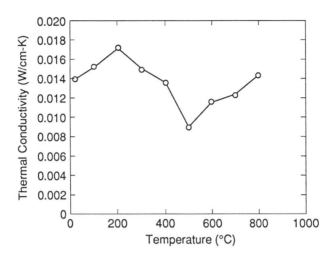

Figure 5. Thermal Conductivity Values at the Pre-Heat Temperature for Carbon-Reinforced Phenolics in the Warp, Ply, Fill and Involute Configurations.

The expansion curves for continuous fiber-reinforced composites usually exhibit considerable hysteresis and residual strain. Generally, these are reduced after several cycles. A thorough study of the length changes experienced by the P75/1962, P75/PEEK and AS4/PES samples has been made. Length changes were measured between -150 and +150°F before and after 10,000 cycles over the same temperature range. The length changes for the non-cycled samples all show large hysteresis and residual changes. These were significantly smaller in magnitude in the extensively cycled samples. This was indicative of the reduced residual strain levels in the laminate. The fabrication residual strains near the reinforcement/matrix interface are partially relieved due to microcracking and microplastic flow in the matrix. Because of the reduced hysteresis, the average CTE obtained by the slope of a line joining thermal strain values at extreme temperatures (-150°F and 150°F) were consistent with the average CTE in the heat (RT→150°F) cool (150°F→-150°F), and heat (-150°F→RT) segments. Also, the average CTE values for each composite were close to the predicted values.

The predicted values were calculated from the Rule of Mixtures relations *(8)*.

• Continuous Fiber Reinforced Longitudinal Thermal Expansion Coefficient:

$$CTE_x = \frac{V_f E_{fx} CTE_{fx} + V_m E_m CTE_m}{E_{fx} V_f + E_m V_m} \tag{5}$$

• Transverse Thermal Expansion Coefficient:

$$CTE_y = CTE_z = CTE_{fy}\sqrt{V_f} + CTE_m\left(1 - \sqrt{V_f}\right)\left(1 + \frac{V_f \nu_m E_{fx}}{E_{fx}V_f + E_m V_m}\right) \quad (6)$$

where E is the modulus and ν is Poisson ratio.

The results before and following 10000 cycles are summarized in Table II.

Table II. Thermal Expansion Results

Material	CTE_x initial	CTE_x cycled	CTE_x predicted	CTE_y initial	CTE^y cycled
P75/1962[0]$_8$	-0.69		-0.60		
P75/1962[0,±45,90]$_s$	-0.45	-0.59	-0.34	-0.14	-0.40
P75/1962[±30,0$_4$]$_s$	-1.02	-1.0		5.83	2.9
P75/PEEK[0,±45,90]$_s$	-0.28	-0.25	-0.30	0.04	-0.20
P75/PEEK[±30,0$_4$]$_s$	-0.39	-0.80	-1.01	10.17	7.6
AS4/PES[0,±45,90]$_s$	1.08	-0.65		1.38	-0.29
AS4/PES[±30,0$_4$]$_s$	-0.66	-0.37		12.6	10.7

Thermophysical Properties for Composites: Literature Data

A data base activity on properties of polymer matrix composites has been underway at CINDAS for some time. Literature data for thermophysical and mechanical properties are compiled, analyzed, and evaluated for entry into the data base. Particular attention is given to factors such as material description (name, composition, reinforcement structure), processing, specimen description (dimensions, orientation), test method (standards, parameters), and material conditioning (moisture effects, aging). In the subsections that immediately follow, several representative sets of data from the database are displayed and discussed.

Thermal Conductivity. Several sets of literature data for thermal conductivity of unidirectional laminates along with a typical epoxy and a pitch-based graphite fiber are shown in Figure 6. The heat flow direction is in all cases along the fiber. Fiber-volume contents are in the range 59-65%. The logarithmic scale accommodates the large range in magnitude of the thermal conductivity observed when comparing an epoxy *(10)*, S-glass/epoxy *(11)*, aramid/epoxy *(12)*, HMS carbon/2002 epoxy *(13)*, GY80 carbon/Code 69 epoxy *(14)*, and a P100 graphite fiber *(15)*, with the increase amounting to a factor of 1000 at room temperature.

Increased thermal conductivity is associated with increasing order and bond strength at the atomic level, and, for carbon fibers, with increased graphite platelet content in the fiber microstructure.

Thermal Expansion. Some published results for thermal expansion of unidirectional carbon/epoxy laminates along with matrix resin and fiber are shown in Figure 7. The data show that the expansion of a unidirectional laminate in the fiber direction (16) is quite similar to the fiber itself (17). On the other hand, for the transverse laminate direction the laminate expansion (18) is midway between the resin alone (18) and the transverse (radial) expansion of the fiber alone (19).

The absorption of moisture by epoxy polymers is widely known to cause swelling of the material. The complication this creates for thermal expansion is demonstrated in Figure 8 by data reported for unidirectional AS/3501-6 carbon/epoxy in the matrix dominant direction (18) along with the 3501-6 epoxy alone (18). The effect of moisture is

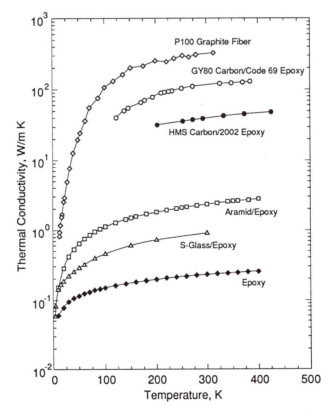

Figure 6. Thermal Conductivity of Various Materials (Literature Data: Unidirectional Composites, Polymer, Fiber).

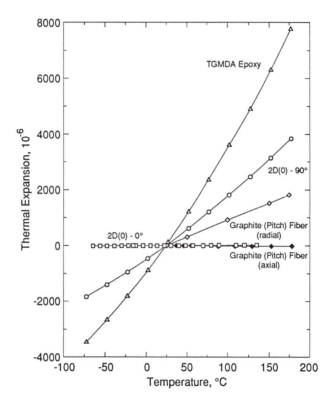

Figure 7.   Thermal   Expansion   of   Carbon/Epoxy   Laminates (Literature Data for Dry Conditions).

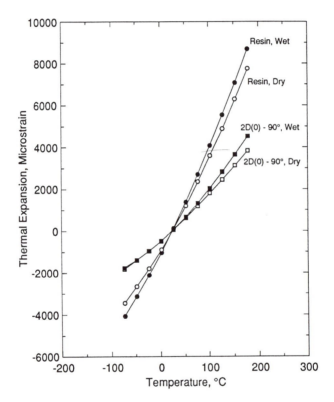

Figure 8.   Thermal Expansion of Carbon/Epoxy Laminates (Literature Data for Wet/Dry Conditions).

to increase the plasticity of the matrix resin, lower the glass-transition temperature, and reduce its stiffness. Clearly, the moisture content of the material, which in the present case is wet (moisture saturated at room temperature) or dry (ambient condition), must be known in order for the data to be meaningful.

Literature Cited.

(1)   Kagan, D.N. in *Compendium of Thermophysical Property Measurement Methods*, Eds., K.D. Maglic, A. Cezairliyan and V.E. Peletsky, Survey of Measurement Techniques, Plenum Publishing, NY, NY, 1984, Vol. 1, Chpt. 12, pp. 461-521.

(2)   Ditmars, D.A., in *Compendium of Thermophysical Property Measurement Methods*, Eds., K.D. Maglic, A. Cezairliyan and V.E. Peletsky, Survey of Measurement Techniques, Plenum Publishing, NY, NY, 1984, Vol 1, Chpt. 13, pp. 527-548.

(3)   Richardson, M.J., in *Compendium of Thermophysical Property Measurement Methods*, Eds., K.D. Maglic, A. Cezairliyan and V.E. Peletsky, Survey of Measurement Techniques, Plenum Publishing, NY, NY 1984, Vol. 1, Chpt 17, pp. 669-685.

(4)   Taylor, R.E., in *Thermal Conductivity 19*, Ed. D.W. Yarbrough, Plenum Publishing, NY, NY, 1988, pp. 403-412.

(5)   Taylor, R.E., Jortner, J. and Groot, H., *Carbon*, **1985**, *23(1)*, pp. 215-222.

(6)   Taylor, Raymond, *High Temp - High Press*, **1983**, *15*, pp. 299-309.

(7)   Rawal, S.P., Misra, M.S. and Wendt, R.G., Report NASA CR-187472, Composite Materials for Space Applications, Martin Marietta Astronautics Group, Denver, CO., Contract NASA1-18230, August, 1990.

(8)   Chawla, K.K., in *Composite Materials, Science and Engineering*, Eds. B. Ilschner and N.J. Grant, Spring-Verlag Publishing, NY, NY, 1987, Chpt. 10.3, pp. 189-191.

(9)   Taylor, R.E., in *Thermal Conductivity 20*, Eds. D.P.H. Hasselman and J.R. Thomas, Jr., Plenum Publishing, NY, NY, 1989, pp. 93-101.

(10) Rosenberg, H.M., in *Nonmetallic Materials and Composites at Low Temperatures*, Eds. G. Hartwig and D. Evans, Plenum Press, NY, NY, 1982, Vol. 2, pp. 311-325.

(11) Hust, J.G., National Bureau of Standards Report NBSIR-84-3003, 1984.

(12) (Bogaard, R.H., HTMIAC/CINDAS, unpublished evaluation of literature data, 1992).

(13) Philpot, K.A. and Randolph, R.E., in *Nonmetallic Materials and Composites at Low Temperatures*, Eds. G. Hartwig and D. Evans, Plenum Press, NY, NY, 1982, pp. 311-325.

(14) McIvor, S.D., et al, *J. Mater. Sci.*, **1990**, *25(7)*, pp. 3127-3132.

(15) Heremans, J., Rahim, I., and Dresselhaus, M.S., *Phys. Rev.*, **1985**, *32*, pp. 6742-6747.

(16) Lauraitis, K.N. and Sandorff, P.E., U.S. Air Force Report AFML-TR-79-4179, 1979, [AD-A085 167].

(17) Yasuda, E., et al., in *18th Biennial Conference on Carbon*, Proceedings and Programs, American Carbon Society, pp. 30-31, 1987.

(18) Cairns, D.S. and Adams, D.F., U.S. Army Report ARO-16370.5-MS, 1981, [AD-A109 131].

(19) Sheaffer, P.M., in *18th Biennial Conference on Carbon*, Proceedings and Programs, American Carbon Society, pp. 20-21, 1987.

RECEIVED April 4, 1995

Chapter 3

High-Temperature Stability of Silicone Polymers and Related Pressure-Sensitive Adhesives

Shaow B. Lin

Specialty Coatings Technology, GE Silicones, 260 Hudson River Road, Mail Drop 1206, Waterford, NY 12188

The thermal stability of five structurally distinctive silicone polymers and the silicone pressure-sensitive adhesives derived therefrom were investigated over a wide temperature range. Thermogravimetric analyses and isothermal weight loss studies showed that the thermal stability of these polymers ranked as follows: polydimethylsiloxane > poly(dimethyl-co-methylvinyl)siloxane > low phenyl poly(dimethyl-co-diphenyl) siloxane > high phenyl poly(dimethyl-co-diphenyl)siloxane > poly (dimethyl-co-diphenyl-co-methylvinyl)siloxane. In addition, cerium and zirconium octoates were shown as effective thermal stabilizers for these polymers and the silicone adhesives at 400 °C and below.

The thermal stability of silicone polymers has been the subject of many studies (1-4) and has been shown to be a function of specific product composition in which these polymers were incorporated (e.g. fillers, additives in silicone elastomers). The stability of many silicone elastomers was also investigated under various testing environments (air, inert atmosphere, or vacuum), temperatures and times. Even though silicones have been long regarded as the material of choice for applications where high temperature resistance is needed, limited information was available on how silicone pressure-sensitive adhesives (PSAs) behave upon high temperature exposure. This study intends to offer a comparison on the thermal stability of those silicone polymers used in silicone pressure-sensitive adhesives, and demonstrate how this relates to the thermal stability characteristics of silicone PSAs.

Commercially significant silicone Pressure-Sensitive Adhesives are well known for their ability to withstand high processing temperatures and harsh environments of modern industrial operations. Premium high-temperature applications are masking tapes for plasma and flame spraying processes which involve an intermittent exposure at 260 °C and beyond. These silicone PSAs typically consist of a siloxane resin and a diorganopolysiloxane polymer, and are prepared by either a blending process or a base-catalyzed chemical reaction process. In general, there are three types of diorganosiloxane polymers in use for making silicone PSAs: polydimethylsiloxane (PDMS) , poly(dimethyl-co-diphenyl)siloxane (PDMDPS) or poly(dimethyl-co-methylphenyl)siloxane, and vinyl-containing silicone polymers (5).

It is widely known that the ratio of siloxane resin to silicone polymer and the

0097–6156/95/0603–0037$12.00/0

type of silicone polymer are two important parameters in the design of silicone PSAs with proper peel and probe tack adhesion properties. However, there are few reports in the literature on the stability of various linear silicone polymers found in typical PSA compositions. Specifically, reports are limited on how the structure of silicone polymer affects the thermal stability of silicone PSAs.

Recently, Lin and Cooper (6) reported that the high temperature stability and the adhesive properties of silicone PSAs can be improved by incorporating in them a small amount of soluble organometallic compounds of rare earth metals or transition metals. The report prompted this investigation into the thermal stability of commercially significant silicone polymers and the adhesives derived therefrom. The effectiveness of selected organometallics as stabilizers was also evaluated.

Experimental

Materials Preparation

Silicone Polymers. Five silicone polymers were selected for the study: one methyl, two phenyls, and two vinyl silicones. The methyl silicone polymer was a dimethylhydroxysiloxy-terminated polydimethylsiloxane (PDMS) gum. The two phenyl silicones were a low phenyl and a high phenyl silicones. These phenyl silicones are dimethylhydroxysiloxy-terminated poly(dimethyl-co-diphenyl)siloxane (PDMDPS) polymers containing 5.3 mole % and 13.6 mole %, diphenylsiloxane units respectively. The two vinyl silicones were: a methyl vinyl and a phenyl vinyl. The methyl vinyl silicone is a poly(dimethyl-co-methylvinyl)siloxane (PDMMVS) polymer having a general structure $M^{vi}D_{\sim 7000}D^{vi}{}_{14}M^{vi}$. The phenyl vinyl silicone is a poly(dimethyl-co-diphenyl-co-methyvinyl)siloxane (PDPMVS) terpolymer having 5.3 mole % diphenylsiloxane units and 0.2 mole % methylvinylsiloxane units. These polymers were prepared via a base-catalysis process and neutralized to slight acidity (about 5 ppm as HCl). The preparation procedures can be found in the literature (4). The structural characteristics and the viscosity of these polymers are summarized in Table I.

Table I. Structure of Silicone Polymers
(General Formula: $MD_xD^{ph}{}_yD^{vi}{}_zM$)

Silicone Type	Methyl D unit, mole %	Phenyl D unit, mole %	Vinyl D unit, mole %	Viscosity, millions cps
PDMS	100	0	0	20.8
PDMMVS	99.8	0	0.2	4.7
PDMDPS1	94.7	5.3	0	62.5
PDMDPS2	86.4	13.6	0	36.7
PDPMVS	94.5	5.3	0.2	117

Silicone Pressure-Sensitive Adhesives. A series of silicone adhesive mixtures were prepared to give a siloxane resin to silicone polymer ratio of 56 to 44 percent by weight. These PSA solutions were prepared by mixing the following components to form a homogeneous mixture: 51.3 parts SR545 siloxane resin solution (a 60 wt.% solids in toluene; GE Silicones), 24.2 parts silicone polymer, and 24.5 parts xylene. The siloxane resin is a copolymer comprising $(CH_3)_3SiO_{1/2}$ (M) units and $SiO_{4/2}$ (Q) units at a molar ratio of about 0.7 to 1, according to ^{29}Si NMR, and about 2 wt.% silanol groups.

The following metallic octoates were acquired from Mooney Chemicals, Inc. for the study: 6 % Cerium Hex-Cem and 6 % Zirconium Hex-Cem; all are metallic 2-ethylhexanoate solutions in mineral spirits.

Thermal Exposures and Characterizations

Materials of interest were placed in aluminum pans and the solvents were removed by drying for about 1 hour @ 170 °C before beginning high temperature exposure testings. Duplicate samples were subjected to 288 °C exposure in a forced-air circulating oven for 3 days. The extent of weight loss was measured daily. Thermogravimetric analysis (TGA) was carried out in air on a Perkin Elmer TGA 7 analyzer. A 12 to 18 mg sample was heated from 30 °C to 630 °C at a rate of 20 °C per minute. The percent weight loss and the rate of weight loss were measured.

The molecular weights and polydispersity of these silicone polymers and a siloxane resin were determined using a Waters Gel Permeation Chromatograph equipped with a set of four ultrastyrogel columns: 10^5 Å, 10^4 Å, 10^3 Å and 500 Å sizes. The chromatograms were collected using a refractive index detector and interpreted based on a set of polystyrene standards.

Adhesive Preparation and Testings

Silicone pressure-sensitive adhesive tapes, having 37.5 to 50 μm adhesive build over 25 μm polyimide film substrate, were prepared directly from the adhesive mixtures. To prepare the peroxide-cured silicone PSA tapes, the adhesive mixtures were catalyzed with 2 wt.% of benzoyl peroxide based on silicone solids. The coated adhesives were first flashed for 90 seconds at 95 °C to remove solvents, then cured for 2 minutes at 170 °C.

The peel adhesion strength of the cured pressure-sensitive adhesives was measured against a clean stainless steel plate, according to ASTM D-1000, on a Scott tester at a pull rate of 30.48 cm per minute at 180 degree angle.

Results and Discussion

Thermal Stability of Silicone Polymers

The thermal-oxidative stability of these silicone polymers at above 260 °C is of particular interest to the design and application of silicone based adhesives. The subject was investigated by two methods: an isothermal weight loss study at 288 °C in air and a kinetic thermogravimetric analysis (TGA) from 30 °C to 630 °C in air.

288 °C Isothermal Weight Loss Study. A 288 °C isothermal exposure to air study was carried out for four types of silicone polymers: PDMS, PDMMVS, PDMDPS2 and PDPMVS. The exposure study included both the as-prepared (control) polymers as well as the cerium- and zirconium- containing systems. The additive containing systems were prepared by adding a 200 ppm equivalence of cerium or zirconium in 2-ethylhexanoate forms to the control polymer. The percent weight loss of these silicone polymers after the exposure is summarized in Table II. A repeat experiment showed that the reproducibility of the test is about 2 wt. %.

As revealed from Table II, significant weight losses were observed in the control materials, even after one day of exposure. The total weight loss of PDMS approximated 29 wt.%. The PDPMVS phenyl vinyl polymer lost its weight completely within 24 hours. Based on the total percent weight loss, the PDMS methyl silicone was the most stable silicone, followed by PDMMVS (43 % wt. loss), then PDMDPS2 (78 % wt. loss), and PDPMVS (100 % wt. loss).

Table II. Thermal Stability of Silicone Polymers (288 °C Exposure to Air)

Silicone Type	Days @ 288 °C	Control, (% wt. loss)	200 ppm Ce, (% wt. loss)	200 ppm Zr, (% wt. loss)
PDMS	1	20.3	12.0	13.0
(Methyl)	2	24.7	16.3	17.4
	3	28.9	18.9	18.8
PDMMVS	1	36.5	16.9	17.8
(Methyl vinyl)	2	40.3	26.3	25.6
	3	42.9	30.2	40.0
PDMDPS2	1	48.8	11.2	13.4
(Phenyl)	2	68.4	15.0	17.4
	3	77.7	17.3	21.3
PDPMVS	1	99.8	9.2	11.4
(Phenyl vinyl)	2	100	11.3	17.4
	3	100	16.7	21.4

Early studies by Patnode and Wilcock (7), Hunter, Hyde and their coworkers (8), and Thomas and Kendrick (9) showed when heated under vacuum, PDMS depolymerized completely to volatile oligomers. Their works also showed that thermal depolymerization through Si-O siloxane bonds was the degradation mechanism for PDMS under vacuum. The degradation products were primarily volatile cyclic mixtures consisting largely of D_3 and D_4, some higher cyclics up to D_{12}, and trace levels of linear monomers including MM and MDM. Lewis (10,11) pointed out the importance of catalyst residues including acids and bases on the degradation of PDMS. Grassie and MacFarlane (3) further pointed out that variations in the volatile composition among previous reports were attributed to the different methods of PDMS polymerization, to the types and levels of impurities, and to the differences in experimental conditions.

According to Thomas and Kendrick (9), the thermal depolymerization of PDMS through Si-O siloxane bonds has an activation energy of 43 Kcal/mol (by isothermal TGA), which is much less than the calculated Si-O siloxane bond energy of 108 Kcal/mol. Thus, the accepted degradation mechanism involves the formation of a transitional structure comprising an intramolecular cyclic siloxane loop, followed by siloxane bond rearrangements leading to the liberation of volatile cyclics. Zeldin, Qian and Choi (12) suggested a transition state involving the weakening of a siloxane bond through localization of the electrons on oxygen by catalyst participation. The concept is consistent with widely known catalytic cleavages of the siloxane bond by species including metal containing electrophiles (e.g. $FeCl_3$ / $(Ch_3CO)_2O$), nucleophiles (e.g. SO_4^{2-}), and strong acids.

Kendrick, Parbhoo and White (13) summarized selected works on the thermal decomposition of PDMS in controlled atmospheres including air, vacuum and argon. The activation energies for thermal-oxidation (in air) process was about 30 Kcal/mol, which is considerably lower than those of the thermal depolymerization processes, 43 Kcal/mol in vacuum and 45-47 Kcal/mol in air. The activation energy for cyclic volatilization, which occurs below 300 °C, is about 14 Kcal/mol in air and 12

Kcal/mol in vacuum. Accordingly, thermal-oxidation process, followed by cyclic volatilization is the energetically favored degradation mechanism for PDMS in air.

This study shows that PDMS has a moderate weight loss during exposure at 288 °C in air. This suggests that unzipped PDMS polymers do not continue chain-reversion through siloxane bonds to lower order cyclics. Instead, they recombine or rearrange to form non-volatile composition. This mechanism also explains the observation that the molecular weight increased with exposure time for silanol-terminated PDMS (3). The presence of oxygen and moisture in air makes this recombination process catalytically favorable. This process is therefore accountable for the moderate weight loss for PDMS in air.

Contrastingly, the PDPMVS silicone polymer decomposed completely in air and no residue was found after exposure. This showed that the main mechanism occurred in PDPMVS during exposure was depolymerization reaction whereby polymers unzipped through Si-O siloxane bonds and reverted completely to volatile species. The absence of non-volatile, higher order cyclics or branched structures suggested that no subsequent rearrangements occurred among the reverted species. Accordingly, the presence of diphenylsiloxane and methylvinylsiloxane units in the PDPMVS siloxane copolymer makes the rearrangement mechanism unfavorable.

The isothermal weight loss profiles of these polymers are shown in Figure 1. It shows that the thermal-oxidative stability relates to the structure of these silicone polymers. Based on these data, two general observations can be made: firstly, the PDMS dimethyl silicone is the most stable in air. Secondly, the presence of substitutes in the silicones including methylvinylsiloxane and diphenylsiloxane units reduce the thermal-oxidative stability.

Kendrick and coworkers (13) summarized that only the depolymerization process was present in a series of substituted siloxane copolymers, including a methylphenylsiloxane containing copolymer. Cyclotrisiloxanes (D'$_3$) and cyclotetrasiloxanes (D'$_4$) were the major degradation products, and the weight loss for all these siloxane copolymers was 100 percent except a trifluoropropyl polymer (75%). The effect of oxygen was attributed to the accelerated production of volatiles and the occurring of crosslinking that led to non-volatile residues. Kendrick's report provides general agreement with this study on the thermal-oxidative stability of siloxane copolymers including PDMMVS, PDMDPS and PDPMVS.

The effectiveness of both zirconium and cerium octoates as thermal stabilizers for these silicone polymers is evident. The extent of weight loss in the stabilized silicone polymers is significantly less than those of the controls, despite the types of silicone. With the exception of the PDMS system, the cerium-containing silicone polymers are slightly more stable than the zirconium-containing counterparts under this condition. In general, the improvement in thermal-oxidation can be achieved by either increasing the activation energy of thermal depolymerization process or catalytically promoting rapid bond rearrangements and/or crosslinking of the unzipped chains. The presence of these metal octoates is likely to promote rapid chain rearrangements and/or crosslinking; therefore it improves the thermal-oxidative stability of these silicone polymers.

In a series of two-hour isothermal weight loss studies in air on a group of vulcanized polysiloxanes between 240 and 340 °C, Critchley et al. (1) observed moderate weight losses in both PDMS and PDMMVS. Polymethylphenylsiloxane showed accelerated weight loss above 320 °C and had the worst stability among the three silicones. Additionally, PDMMVS had poorer stability than PDMS at temperatures below 310 °C. While specific conditions differ, the general trend appears to be in good agreement with the findings in the study.

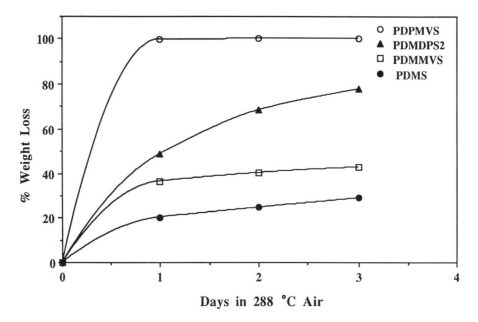

Figure 1. Percent weight loss in silicone polymers after 288 °C isothermal exposure to air: (●) PDMS, (□) PDMMVS, (▲) PDMDPS2, and (○) PDPMVS.

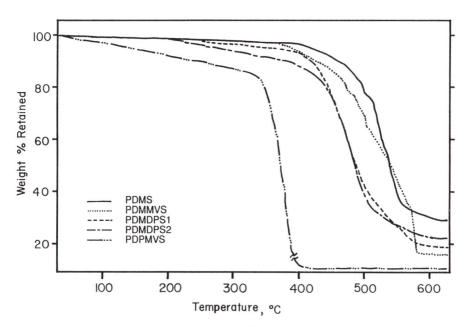

Figure 2. Thermogravimetric analysis of silicone polymers: PDMS, PDMMVS, PDMDPS1, PDMDPS2, and PDPMVS.

Thermogravimetric Analysis of Silicone Polymers. The thermal stability of the silicone polymers were compared over a wide temperature range using a Perkin Elmer TGA 7 thermogravimetric analyzer. The silicone polymer was placed in a platinum pan heated from 30 °C to 630 °C in air at a rate of 20 °C per minute. The weight loss versus temperature profiles are shown in Figure 2.

The figure indicates that substantial weight losses occurred on all these silicones approaching 600 °C. Only a moderate amount of weight loss was observed at 400 °C and below, with the exception of PDPMVS phenyl vinyl silicone. Based on the amount of the weight loss at 400 °C, the thermal stability of these silicone polymers ranked as follows: PDMS (-3.8 wt.% @ 400 °C) > PDMMVS (-6.2 wt.%) > PDMDPS1 (-6.8 wt.%) > PDMDPS2 (-11.7 wt.%) > PDPMVS (-89.6 wt.%). The polydimethylsiloxane was the most stable polymer, followed by the methyl vinyl siloxane, then the low phenyl siloxane, the high phenyl siloxane, and the phenyl vinyl siloxane.

At 400 °C and below, no charring through Si-C and C-H bonds occurred (1); instead, cyclic silicones were the primary products of thermal degradation through Si-O bond scissions and rearrangements. Therefore, the extent of weight loss is indicative of the tendency of such thermal degradation through the Si-O siloxane bonds and the ability of forming a stable composition through rapid rearrangements.

As shown in Figure 2, thermal degradation was complete around 600 °C and the silica char weighed in the following order: PDMS (29 wt.% remained), PDMDPS2 (22 wt.%), PDMDPS1 (19 wt.%), PDMMVS (16 wt.%), and PDPMVS (0 wt.%). The composition of the silica char was not analyzed. However, the weight of the residue is an indication of the extent of "silicanization" that occurred. Obviously, a complete reversion reaction took place in PDPMVS. This was also observed in the isothermal weight loss study.

Thermogravimetric Analysis of Stabilizer-Containing Silicone Polymers. The effect of cerium and zirconium octoates as stabilizers in silicone polymers over a wide temperature range was investigated. Metallic octoates at an amount of 200 ppm cerium or zirconium equivalent were added to silicone polymers. The TGA weight loss profiles for PDMDPS2 high phenyl silicone are shown in Figure 3. Based on the percent weight loss at 400 °C, the stability of these materials ranked: Ce-stabilized (-5.3 wt.%) > Zr-stabilized (-6.8 wt.%) > Control (-11.7 wt.%). The data suggested that both additives succeeded in retarding the thermal degradation.

The non-volatile residues weighed at the end of the testing showed that the control had 22 wt.% remaining, Ce-stabilized had 33 wt.%, and Zr-stabilized 46 wt.%. Presumably, extensive silicanization and crosslinking took place within the silicone polymers when under the influence of zirconium.

In Figure 4, the rate of weight loss versus temperature profiles for these three materials are shown. These are the first order derivative curves of the TGA thermograms shown in Figure 3. The control PDMDPS2 starts degradation around 380 °C and the rate of weight loss maximized around 472 °C. By comparison, the cerium- and zirconium- stabilized materials did not allow significant degradation until around 450 °C, and the rate of weight loss maximized around 504 °C, and 523 °C, respectively.

Similar observations were made on the PDPMVS silicone polymer system. The stabilized system exhibited only moderate weight loss at 400 °C (Zr: -3.9 wt.%; Ce: -4.8 wt.%) as opposed to a loss of 89.6 wt.% for the control. Additionally, the non-volatile residues weighed as follows: the Zr-stabilized (36.5 wt.%), the Ce-stabilized (22.3 wt.%), and the control (0 wt.%).

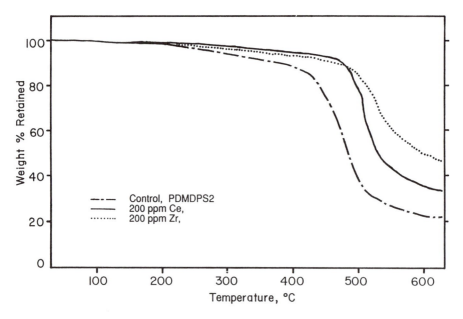

Figure 3. Thermogravimetric analysis of PDMDPS2 phenyl silicone polymer: (a) Control, (b) 200 ppm cerium stabilized, and (c) 200 ppm zirconium stabilized.

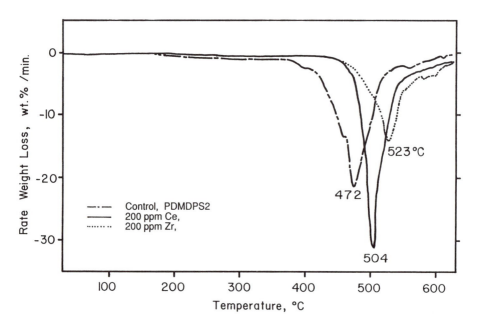

Figure 4. Rate of weight loss per thermogravimetric analysis in air for PDMDPS2 phenyl silicone polymer: (a) Control, (b) 200 ppm cerium stabilized, and (c) 200 ppm zirconium stabilized.

Molecular Weight Characterization. In an attempt to illustrate the structural changes that occurred upon thermal oxidation, PDMS silicone polymer and a MQ siloxane resin were subjected to 8 hours of continuous exposure to air in a 260 °C oven. The stabilized materials contained 200 ppm cerium as additive. The exposed materials were then redissolved in toluene for GPC separation analysis. The GPC chromatograms for the PDMS silicone, before and after exposure, and the cerium-containing PDMS are shown in Figure 5. The molecular weights and polydispersity of the materials, before and after exposure, are shown in Table III.

Table III. GPC Analysis of Aged Silicone and Siloxane Resin

Material	Exposure to air, 8 hrs @ 260 °C	M_w	M_n	Polydispersity
PDMS	Before	630K	383K	1.65
(control)	After	388.6K	140K	2.77
PDMS	Before	636K	389K	1.63
(stabilized)	After	651K	355K	1.83
MQ Resin	Before	3698	2210	1.67
(control)	After	3756	2322	1.62
MQ Resin	Before	3929	2543	1.54
(stabilized)	After	4021	2379	1.69

As exhibited, the molecular weights (M_w and M_n) of the control MQ siloxane resin remained substantially unchanged after exposure. The MQ resin exhibited reasonable thermal stability. The cerium stabilized MQ siloxane resin showed minimal changes in both the molecular weights and polydispersity. Further thermal exposure of these resinous materials for a total of 24 hrs led to insoluble mass and moderate weight losses. This suggests that the changes are due to volatilization of low molecular weight species and further crosslinking of resin molecules, through silanol condensation.

Contrarily, the molecular weight of the PDMS silicone reduced substantially after exposure. The exposed, control PDMS showed an increase in polydispersity and a reduction in molecular weights (M_w and M_n). This is direct evidence of depolymerization of PDMS through chain unzipping, leading to the reduction in molecular weights and the broadening of polydispersity. The presence of cerium octoate in PDMS allowed only small changes in both molecular weights and polydispersity.

Thermal Stability of Silicone Adhesives

Thermogravimetric Analysis of Silicone Adhesives. The thermal degradation kinetics of a cured, PDMS-type methyl silicone PSA was monitored using a Perkin Elmer thermogravimetric analyzer. The cured silicone PSA was prepared by catalyzing the PDMS-containing adhesive mixture with 2.0 wt.% benzoyl peroxide. About 15 mg of this cured adhesive was subjected to a three-day isothermal air exposure at 288 °C in the TGA analyzer. The thermogravimetric weight loss profile is shown in Figure 6. Further treatment of the weight loss profile of this methyl silicone PSA gave a semi-logarithmic equation with excellent correlation shown as follows:

% weight loss = -6.144 + 7.19 log(exposure minutes)

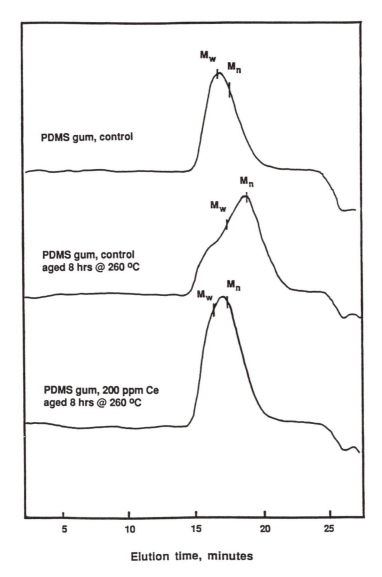

Figure 5. Gel Permeation Chromatograms of PDMS silicone polymer: (a) before and (b) after 8-hour air exposure at 260 °C, and (c) the cerium-containing PDMS after the exposure.

It appears that the thermal-oxidative degradation followed first-order kinetics at loss rates of 12.4 % by weight in the first 6.3 hours, 16.8 % within the first 24 hours, and a total of 19.6 % after 3 days exposure at 288 °C.

Isothermal Aging of Silicone Adhesives. Silicone adhesive solutions comprising of 56 parts of a MQ siloxane resin and 44 parts of a silicone polymer were prepared according to the method previously described. An amount equivalent to 200 ppm as metal in metallic octoate solution was dispersed homogeneously into the adhesive mixtures. These adhesives were thoroughly dried at 152 °C before subjecting to the 288 °C exposure in air. The percent weight loss in each adhesive is reported in Table IV. Under the same exposure condition, the neat MQ siloxane resin lost about 5 wt.%.

Table IV. Thermal Stability of Silicone Adhesives

Adhesive Type	Days @ 288 °C	Control, (% wt. loss)	200 ppm Ce, (% wt. loss)	200 ppm Zr, (% wt. loss)
Methylvinyl PSA	1	16.6	18.9	16.9
(PDMMVS)	2	18.0	21.2	20.1
	3	20.9	23.3	21.7
Phenyl PSA	1	17.0	19.4	17.5
(PDMDPS2)	2	19.2	22.0	20.1
	3	20.7	23.3	22.1
Phenylvinyl PSA	1	18.0	19.0	17.4
(PDPMVS)	2	21.5	20.8	19.5
	3	24.0	22.4	21.6

As exhibited, these adhesives experienced only moderate weight losses during exposure which greatly contrasted with the respective silicone polymers. For example, PDMDPS2 phenyl silicone lost about 77 % by weight after 3 days of exposure (shown in Table II), whereas the phenyl PSA lost 20.7 % by weight under the same condition. This indicated that the phenyl polymer without stabilizers lost only about 17.9 wt.% in an adhesive mixture comprising of 56 parts of the MQ siloxane resin and 44 parts of the phenyl gum. The PDPMVS phenylvinyl gum of the adhesive mixture showed a loss of 21.2 wt.%, instead of a complete degradation. The presence of the MQ siloxane resin is credited for the thermal stability of the gums without stabilizers. A plausible mechanism is when the siloxane resin interacts or reacts, for example through silanol groups of the MQ resin, with unzipped silicone chains resulting in the formation of stable branched structure. Therefore, these stabilizers did not appear to impart additional stability to the adhesive matrices.

It was further observed that after exposure, the aged, control adhesives became dried granules, the zirconium stabilized adhesives left a residue mass of clear hard pieces, whereas the cerium stabilized adhesive residue was a clear, smooth continuous mass. The significance of the residue appearance is yet to be determined.

The presence of a "structured" resin-polymer phase could be responsible for the improved thermal stability of silicone PSAs over that of a simple blend system. Based on the dynamic mechanical study on silicone PSAs, Copley (14, 15) suggested that both PDMS and PDMDPS silicone PSAs had a two-phase morphology: a small yet detectable silicone gum phase and a dominant resin-silicone polymer coexisting phase. Based on this morphological model, the silanol groups of the siloxane resin are readily available for rearrangement reactions with silicone polymers during exposure. It therefore, minimizes the production of volatile cyclic siloxanes.

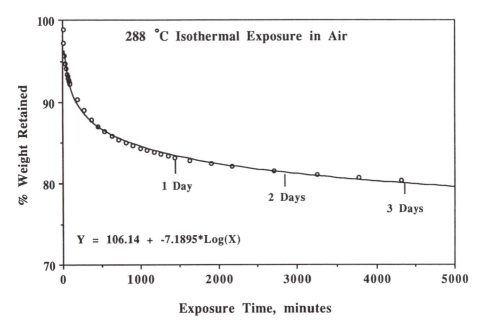

Figure 6. A 3-day 288 °C isothermal thermogravimetric analysis of a benzoyl peroxide-cured PDMS methyl silicone adhesive.

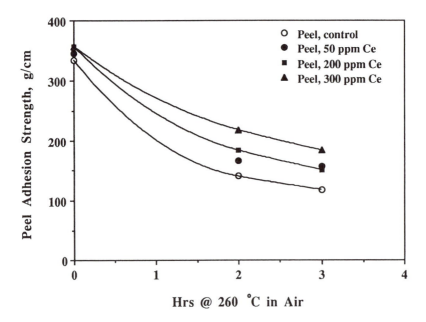

Figure 7. Thermal aging effect on the peel adhesion strength of a PDMS-based silicone pressure-sensitive adhesive.

Thermal Aging Effect on Silicone Adhesive Properties

A series of PDMS methyl silicone PSA solutions having varying levels of cerium additive was prepared, then catalyzed with 2 wt.% benzoyl peroxide. The cured silicone pressure-sensitive adhesive tapes of 2.543 cm width by 30.48 cm length, with the adhesive side exposed to air, were placed for various periods of time in a forced air circulating oven @ 260 °C. The peel and probe tack adhesion properties were measured before and after the thermal aging. Summarized in Table V are the peel adhesion properties of these adhesives after the high temperature aging test.

Table V. Thermal Aging of Methyl Silicone Adhesives

ppm Ce	Peel, initial (g / cm)	Peel, after 2 hrs (g / cm)	Peel, after 3 hrs (g / cm)	Tape Appearance, after exposure (3 days, 260 °C)
0	335	140	117	Severe curl
50	346	167	156	Moderate curl
200	357	184	151	Slight curl
300	357	218	184	None

The peel adhesion profiles of the thermally aged silicone PSAs are illustrated in Figures 7. As shown, silicone PSAs lose pressure-sensitive peel adhesion upon thermal exposure. The stabilized adhesives exhibited a better retention of the adhesive properties than the control adhesive, and the effectiveness appears to be proportional to the cerium concentration.

The appearance of the 72-hour aged adhesive tapes is also reported in Table V. The control adhesive tape severely curled after exposure. The extent of the tape deformation seemed to correlate well with the cerium concentration. Although the stabilization mechanism was not analytically investigated, the effect of cerium octoate as a stabilizer for PDMS silicone adhesives was evident. This deformation is attributed to increased crosslinking and hardening of the adhesive matrix.

While the isothermal weight loss study suggested the cerium and zirconium based additives offer little further weight retention, it is clear that the adhesive properties of the stabilized silicone PSAs were better retained during high temperature air exposure. It is viewed that the presence of such stabilizers deterred the undesirable crosslinking and hardening of the adhesive composition and therefore, prevented the loss of adhesive characteristics.

Summary

The thermal stability of five structurally distinctive silicone polymers and a series of silicone pressure-sensitive adhesives derived therefrom was investigated over a wide temperature range. The silicone polymers included a polydimethylsiloxane (PDMS) methyl silicone, two poly(dimethyl-co-diphenyl)siloxanes having low (PDMDPS1) and high phenyl (PDMDPS2) contents, a poly(dimethyl-co-methylvinyl)siloxane (PDMMVS) methyl vinyl silicone, and a poly(dimethyl-co-diphenyl-co-methylvinyl)siloxane (PDPMVS) phenyl vinyl silicone. Additionally, the effectiveness of cerium and zirconium octoates as stabilizers for these polymers was studied.

Thermogravimetric analysis showed that these silicone polymers underwent significant thermal degradation in air at high temperatures. Based on the extent of weight loss at 400 °C, the relative thermal stability of these polymers ranked as follows: PDMS > PDMMVS > PDMDPS1 > PDMDPS2 > PDPMVS. A series of 3-

day isothermal oxidative study at 288 °C showed the weight loss for these polymers ranked in the same order, ranging from a moderate 29 wt.% for PDMS to 100 wt.% for PDPMVS. A previously proposed thermal oxidation process provides an explanation for the contrasting stability among these polymers. The degradation mechanism involves thermal depolymerization through Si-O bonds, unzipped chain rearrangements, and cyclics volatilization. The presence of methylvinylsiloxane and diphenylsiloxane units may have impaired the ability for rapid rearrangements among the unzipped chains during the oxidation.

According to the TGA analysis and the 288 °C isothermal weight loss studies, cerium and zirconium octoates were effective in improving the thermal stability of these silicone polymers. Only moderate weight losses were found in these silicones including diphenylsiloxane- and methylvinylsiloxane-containing copolymers after the 3-day 288 °C isothermal exposure. The presence of these metal octoates deterred the rapid degradation tendency of these copolymers.

The thermal stability of the silicone PSAs was inherently better than their respective silicone polymer components. None of the prepared silicone adhesive mixtures experienced more than a moderate weight loss. This inherent thermal stability of silicone PSAs was credited to the presence of a silanol-bearing MQ siloxane resin, which reacted with unzipping polymers and resulted in a non-volatile, crosslinked adhesive matrix.

Based on the weight loss comparison, metal octoates do not add to the thermal stability of silicone PSAs including methyl vinyl, phenyl and phenyl vinyl systems. Upon isothermal exposure, the pressure-sensitive peel adhesion strength of these silicone adhesives decreased with time. Yet, the results showed that cerium octoate was an effective stabilizer for silicone PSAs in deterring the irreversible loss of the adhesive property.

Acknowledgment

The author would like to thank Robert Rowland for technical support, Sue Kelly for TGA analysis, Gerald Dudding and William Cooper for GPC analysis. The support of GE Silicones management is greatly appreciated.

References

1. Critchley, J.P.; Knight, G.J.; Wright W.W., *"Heat Resistant Polymers"*, Plenum Press, New York and London, **1983**; Chapter 6, pp. 323-361.
2. Goldovskii, E.A.; Kuzminshill, A.S., *Polym. Sci. Tech.*, **1979**, *6(4)*, p.75.
3. Grassie, N.; MacFarlane, I.G.; *Eur. Polym. J.*, **1978**, *14*, p.875.
4. Aseyeva, R.M.; Mezhikovskii, S.M.; Kholodovskaya, A.A.; Selskaya, O.G.; Berlin, A.A.; *Polym. Sci. U.S.S.R.*, **1973**, *A 15* (8), p. 2104.
5. Satas D.(Ed.);*" Handbook of Pressure-sensitive Adhesive Technology,"* **1982**, Van Nostrand Reinhold; p. 344.
6. Lin, S.B.; Cooper. W.B.; *US Patent* application (5/92).
7. Patnode, W. and Wilcock, D.F.; *J. Am. Chem. Soc.*, **1946**, *68*, p. 358.
8. Hunter,M.J.; Hyde, J.F.; Warrick, W.L.;Fletcher, H.J.; *J. Am. Chem. Soc.*, **1946**, *68*, p. 667.
9. Thomas, T.H.;Kendrick, T.C., *J. Polym. Sci.*, Part A-2, **1969**, *7*, p. 537.
10. Lewis, C.W.; *J. Polym. Sci.*, **1958**, *33*, p. 153.
11. Lewis, C.W.; *J. Polym. Sci.*, **1959**, *337* p. 425.
12. Zeldin, M.; Qian, B.; Choi, S., *J. Polym. Sci., Polym. Chem.*, **1983**, *21*, p. 1361.

13. Kendrick, T.C.; Parbhoo, B.; White, J.W. *" The Chemistry of Organic Silicon Compounds,"* Patai, S. and Rappoport, Z. (Eds.),John Wiley & Sons Ltd., **1989**, Chapter 21, pp.1289 - 1361.
14. Copley, B.C., *ACS Organic Coatings & Appl. Polym. Sci.*, **1983**, _48_, pp. 121-125.
15. Copley, B.C., M.S. Thesis, Univ. Minnestota, **1984**.

RECEIVED February 17, 1995

Chapter 4

Testing Materials at Elevated Temperatures Using Moiré Interferometry

Robert Czarnek

Concurrent Technologies Corporation, 1450 Scalp Avenue, Johnstown, PA 15904

New developments in moiré interferometry have expanded its applicability to high-temperature regimes. Basic principles of moiré interferometry are discussed, then its application to testing materials at elevated temperatures is presented. An achromatic interferometer with extended working distance allows routine measurements at atmospheric pressures at temperatures of order of 200°C. Remote control of this interferometer allows its implementation in a vacuum oven. Successful measurements at full standard sensitivity of 417nm per fringe order were performed at temperatures near 1000°C. A few examples of the application of this new technique are presented. New specimen preparation techniques are discussed: methods commonly used in microelectronics industry were adapted to produce a near zero thickness diffraction gratings directly on the surface of the specimen. Such a grating was successfully tested for survivability at temperatures up to 1100°C.

In recent years many new materials have been developed for applications in high temperature environments. However, their use in high performance structures has been limited by the lack of a sufficient database related to their mechanical properties. Most of the existing test methods were developed for studying homogeneous materials at ambient temperatures. Some were adapted to high temperature applications. Unfortunately, these traditional techniques rely on the use of extensometers or resistance strain gauges, and may provide incomplete or erroneous data when applied to fiber reinforced composites or other materials where the properties vary. The major reason for the errors in such measurements is the inability to detect nonuniformities in the strain field. These effects are usually random, and as such are impossible to predict. Statistical methods of compensation for these variations are often used, but can be very misleading and may mask other sources of error, including imperfections in the fixturing, and assumptions made about the material behavior. Furthermore, the point methods of strain measurement do not provide any information about localized defects or failure initiation.

0097–6156/95/0603–0052$12.00/0

The problems mentioned above can be avoided if full-field measuring techniques are used. The most common full-field methods are the grid method, speckle photography, holography, speckle interferometry, and moiré interferometry. The first two were developed for lower sensitivities and are practical in measurements of large deformations. The last three, representing basically the same family of interferometric methods, offer much higher sensitivity.

All the above-mentioned techniques have been tried in high temperature deformation measurements with varying degrees of success. The major obstacle in the implementation of interferometric methods is their sensitivity to not only measured deformation, but also to vibrations in the surrounding equipment and to the distortions introduced by atmospheric thermal currents. A successful approach to overcoming these obstacles and implementing moiré interferometry at elevated temperatures is presented here.

High-Sensitivity Moiré Interferometry

High-sensitivity moiré interferometry is a method of precisely measuring the deformation of solid bodies under various loading conditions. It is a full-field technique that can be classified as a combination of holography and the traditional moiré method. It maintains all the sensitivity of holography and at the same time offers the easy-to-interpret interferograms associated with traditional moiré methods. The contrast of the patterns is usually better than in either one of the parent techniques.

High-sensitivity moiré interferometry was introduced by Guild in 1956 as a powerful experimental method of measuring deformations of solid bodies (1). It extends the sensitivity of traditional moiré interferometry by two orders of magnitude. Since its introduction, a wide range of materials has been tested by moiré interferometry, including simple isotropic materials and high-performance composites. Almost all experiments described in the literature, however, have been performed at room temperature. Some attempts at making unidirectional measurements at higher temperatures have been made , but the instabilities of the systems used, the high cost, and the complexity of the experimental work within a very limited temperature range prohibited these procedures from being widely used.

Moiré interferometry is an optical measuring technique based on the interference of two beams of light (2). Figure 1 illustrates the basic features of a moiré interferometry system. Two collimated and mutually coherent beams of light, A and B, illuminate a diffraction grating attached to a flat surface of the specimen. The grating follows the deformation of the specimen surface and modulates the shape of the wave fronts of the diffracted beams. The first-order diffracted beam A' of incident beam A, and B', from incident beam B interfere and produce an interference fringe pattern that is recorded by the camera.

At the initial no-load configuration, the two diffracted beams emerge along the normal to the specimen surface and the pattern produced is a uniform intensity field called a "null" field. After a load is applied, the specimen deforms and the new pattern generated is a contour map of the phase difference of the two interfering beams. The fringes are assigned numbers called fringe-orders, defined by Equation (1).

Since a fringe order is directly proportional to the in-plane displacements of points on the specimen surface, in a direction perpendicular to the grating lines, the new fringe pattern is a contour-map of this component of the displacement field. Equation (2) defines the relationship between the fringe-order and the displacement it represents for the case presented above.

There are numerous ways to implement the basic idea of moiré interferometry. Usually, a collimated beam of coherent light is divided into two or four parts and directed toward the specimen grating at the proper angles. The whole system is built on a large and heavy holographic table insulated from vibrations. In such laboratory systems motion of any optical element affects the results. In addition the presence of a hot body in the vicinity of such a system could introduce unacceptable and inestimable errors because of induced air currents.

$$N = \frac{S}{\lambda} + C \tag{1}$$

where N   is the fringe order
          S   is the phase difference in terms of the distance between the wave fronts
          λ is the wavelength of light
          C   is a constant corresponding to rigid body motion

$$U = \frac{N}{2 f_s} \tag{2}$$

where U   is the displacement
          f_s is the frequency of the specimen grating

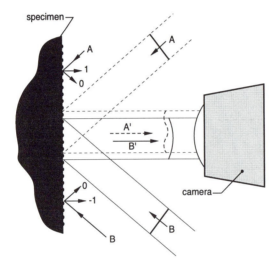

Figure 1.   Basic moiré interferometry system.

The influence of air currents can be reduced by minimizing the distance between the paths of the illuminating beams. This solution is effective for small specimens at moderate temperature levels. However, for larger specimens at higher temperatures, atmospheric thermal currents produce unstable interference patterns, which cannot be used as reliable sources of data even if short photographic exposure times are employed. Furthermore, the use of a window, necessary at higher temperatures at normal atmospheric pressure, introduces to the illuminating beams additional distortions that are hard to estimate.

The application of conventional interferometry at temperatures above 250°C cannot be accomplished without risk of significant error. The achromatic interferometer, developed by the author for use in a materials-testing laboratory (3-4) is much more resistant to both vibrations and atmospheric thermal currents than the traditional moiré interferometers and was, for this reason, chosen for high-temperature work. It was modified to maximize the distance between the specimen and the optics while preserving the compactness of the configuration.

Achromatic, High-Temperature Interferometer

The high-temperature achromatic interferometer is schematically illustrated in Figure 2 (5-6). A collimated beam illuminates a diffraction grating, called compensator, at normal incidence. The two first order diffracted beams are directed by a system of four mirrors so they illuminate the specimen grating at precisely defined angles. In a vacuum environment, one of the mirrors can be adjusted with high-precision, remotely operated, electric actuators. Both diffraction gratings, the compensator and the specimen grating had a frequency of 1200 lines per millimeter. This frequency defines the sensitivity of measurements. In all the experiments presented in this paper this sensitivity was 417 nanometers per fringe order and resolution better than 100nm.

At higher temperatures a vacuum is necessary to protect the instrument from thermal deformation and prevent the motion of the interference fringes associated with air currents. The small sensitivity to the movement of the illuminating beam allows the laser and the collimating optics to be positioned outside the chamber, thereby reducing the requirements for the size of the oven.

The literature shows that several attempts have been made in the past to eliminate the distortions and errors associated with thermal currents. Solutions such as short exposure times and even immersion in fluids were tried. Unfortunately the only improvement accomplished was the capability to record the pattern on a film. The errors were neglected and impossible to estimate.

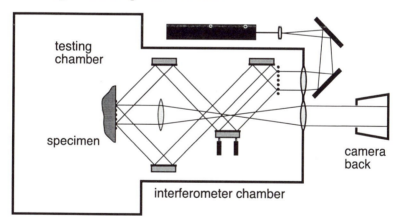

Figure 2.   Moiré interferometer for high temperature applications. (Figure reprinted courtesy of the Society for Experimental Mechanics.)

To obtain all three components of the two-dimensional strain tensor it is necessary to measure two orthogonal components of the in-plane displacement field. To achieve this, two systems were built on a common frame, creating a stable and compact instrument. A total of four remotely controlled actuators were used to control this system. Their configuration allows not only fine tuning of the interferometer to eliminate fringes in the initial field, but also to introduce rigid-body rotation of the whole instrument with respect to the specimen to form a carrier pattern of rotation. Since the carrier pattern is a contour map of a rigid body motion it does not affect the information about the strain distribution on the specimen surface. It provides more and conveniently positioned data points during the data collection. This is important for high resolution analysis of deformation of the specimen (7).

The prototype oven illustrated in Figure 2 is constructed from two large-diameter (450mm) steel pipes welded together to form a "T". This configuration minimizes the volume to be handled by the vacuum system while providing a relatively large space for the specimen and the interferometer loading frame. The specimen compartment is 750 mm high; removable plates at each end allow easy access to the specimen and fixturing. The whole apparatus is mounted to a steel frame suspended on four air mounts and stabilized by precision leveling valves. The incident light source is supplied by a laser located outside the vacuum chamber. The windows are designed as single lenses and act as collimating elements in the light path. The interferometer can tolerate imperfect collimation of the incident beam so that distortions produced by the window elements are not critical.

The temperature of the interferometer and the specimen are monitored by thermocouples connected to a data acquisition system.

High-Temperature Specimen Grating

Before any measurement can be made using moiré interferometry a diffraction grating must be prepared on the specimen surface. The standard procedure, illustrated in Figure 3, is to replicate the grating from a special mold. The mold is a phase diffraction grating treated with a parting substance and coated with a thin metallic layer by vapor deposition. The mold is cemented to the specimen surface with a two-component adhesive and pulled away after the adhesive is cured. The thin metallic layer remains on the surface of the specimen, providing high reflectivity to the grating. When epoxy resin and aluminum coating are used, such a specimen grating can

Thin layer of metal is vacuum deposited on the surface of a phase diffraction grating treated with a parting agent.

The mold is cemented to the specimen surface.

The mold is separated from the specimen. A reflective phase diffraction grating is replicated on the specimen surface.

Figure 3.   Preparation of a specimen grating using replication process.

survive temperatures up to about 220°C. This is sufficient for most applications involving polymer matrix composites. More work is needed to develop a replication procedure for a new family of high-temperature polymer matrix composites that can be used at temperatures as high as 450°C. Unfortunately, most high-temperature epoxy adhesives require curing at elevated temperatures, making it difficult to use them for replication.

High temperature silicon rubbers can be used to produce specimen gratings that can survive up to about 350°C. Above this temperature they deteriorate very quickly. The replication process for silicone rubbers is very similar to that used for epoxy except the mold is not coated with aluminum. The replicated specimen grating is often used uncoated, or if the exposure time is critical, it is coated after replication. In such a case gold is preferred as the coating material. Aluminum can be applied but unless the rubber is perfectly cured and degassed in a vacuum for several hours it turns black instead of becoming reflective.

Since the long-range goal of this program was to provide the capability to test various materials at temperatures as high as 1000°C, a new method of specimen grating preparation was needed. A procedure similar to the one used in the micro-chip industry was chosen since earlier work based on this technique had demonstrated that it would be successful (8).

As illustrated in Figure 4, these high-temperature gratings are produced by etching a thin metallic layer through a photoresist mask. First, the surface of the specimen must be polished so that its roughness is on the same order of magnitude as the wavelength of the light source used to form the diffraction pattern. The specimen is then cleaned and degreased using solvents, and positioned in a vacuum chamber. A surface layer of atoms is removed with an ion beam prior to coating with a thin layer of gold by evaporation. After removal from the vacuum chamber, a thin layer of photoresist is applied to the specimen surface.

The specimen is then positioned in an interferometer that uses two-beam interference to produce uniformly spaced planes of constructive interference separated by planes of destructive interference. A light source, such as a krypton laser at wavelength of 413 nm, may be used to generate uniformly spaced planes of interference at a high frequency. When the photoresist is exposed and developed it creates a mask consisting of thin, parallel and uniformly distributed strips. At this point the specimen is ready for etching, which can be done using a plasma source or a chemical bath. Once a satisfactory result is attained, the photoresist is washed out and the remaining gold strips become an amplitude diffraction grating.

A simplified procedure published by Kearney and Forno (9) can be used as an alternative. It produces gratings slightly less resistant to temperatures but does not require ion etching therefore reducing the cost and effort required to prepare a specimen.

Measurements at High Temperature

The initial tests were performed on graphite/epoxy composite specimens subjected to thermal loading only. The specimen was prepared by replicating a grating from a holographic mold using an epoxy cement that cures at room temperature. The sequence of the plies was $[0_2/60_2/-60_2]_s$, and the grating was replicated on the edge of the plate. The deformation was measured in steps up to 180°C. The corresponding set of interference patterns representing different stages of the horizontal displacement field is presented in Figure 5. The deformation is due to the edge effect common to composites and to the release of residual stresses near the glass-ransition temperature for the matrix material.

The same process was used to measure thermal deformation of a carbide-tipped cutting tool. The vertical displacement field, recorded at 219°C, is presented in Figure 6. Clearly visible in this pattern is the difference in the coefficients of thermal expansion for steel and tungsten carbide. High levels of shear strain can also be noticed at the interfaces in the vicinity of singularity points.

A silicon rubber grating was tested for survivability on a block of steel. Good quality fringe patterns were recorded at temperatures up to about 350°C. Above this temperature silicone rubbers became chemically unstable. The highest temperature at which measurements were successfully taken was 357°C. A fragment of the corresponding horizontal displacement pattern is illustrated in Figure 7. However, soon after the exposures were taken the grating deteriorated and the patterns gradually vanished.

A number of experiments have been performed to validate the concepts of the vacuum interferometer and plasma-etched grating. The first was conducted on a block of silicon-carbide ceramic. The high temperature diffraction grating was tested for survivability and diffraction at elevated temperatures. The specimen was positioned in a furnace and heated in 100C° increments to 1100°C. At each temperature step the grating was illuminated with a laser beam and a diffracted beam was observed on a screen. No deterioration of the grating was noticed and the diffraction pattern was of good quality at each step, indicating that moiré measurements could be made at these elevated temperatures. This specimen was then wrapped with a resistance wire and positioned in the vacuum oven. The specimen was heated to above 520°C, and a good quality pattern was recorded.

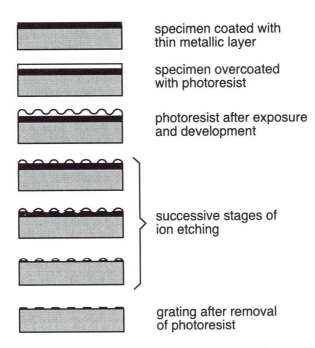

specimen coated with thin metallic layer

specimen overcoated with photoresist

photoresist after exposure and development

successive stages of ion etching

grating after removal of photoresist

Figure 4.   Preparation of a high-temperature specimen grating.

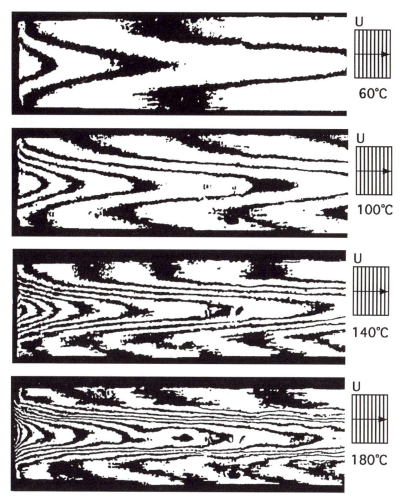

Figure 5.   Thermal deformation at the edge of a graphite/epoxy specimen.

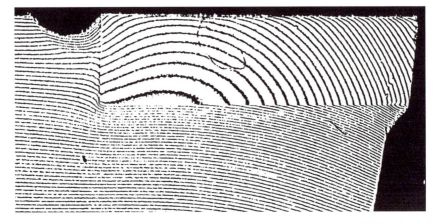

Figure 6. Vertical displacement pattern representing thermal deformation of a carbide tipped cutting tool at 219°C.

Figure 7. Horizontal displacement patern representing thermal deformation of a steel block prepared with a silicon grating and coated with gold. The specimen temperature was 357°C.

Figure 8. Moiré pattern representing the horizontal displacement field due to thermal deformation of quartz specimen at 986°C. (Figure reprinted courtesy of the Society for Experimental Mechanics.)

A successful experiment has also been performed at 986°C. A quartz plate specimen was cemented to a ceramic heater. No mechanical load was applied but the difference in thermal expansion of the quartz and the underlying ceramic material caused the development of stresses high enough to fracture the quartz plate. Figure 8 demonstrates a fragment of a fringe pattern recorded at this temperature. As can be seen, the contrast is very good and the fringes are stable and well defined. In this experiment, gold was used for the diffraction grating, but 1000°C is a practical maximum for this metal. For higher temperatures, experiments with alternate metals such as platinum or tungsten are required. Alternatively, etching the grooves directly into the substrate may increase the stability of the grating and allow further increases in temperature. However, since different materials behave differently during the etching process a metallic coating provides much better success rate without requiring extra research for every tested material.

The purpose of the experiments performed on quartz and silicon-carbide specimens was to demonstrate the practical temperature limits for moiré interferometry measurements. The specimens were purchased polished and flat, therefore requiring minimum effort in specimen preparation. They proved that a high-temperature grating can be produced on real structural materials as long as the surface can be polished to optical quality, sufficient to reflect an image without diffusing the illuminating light. Most metals, ceramics and plastics satisfy this requirement.

The next experiment demonstrated the capability of high-temperature moiré interferometry to handle real structural materials. Metal matrix composite coupon was chosen as one of the most difficult to handle. Due to the technology used in manufacturing their surface usually deviates from flatness and has some texture. The surface of the coupon was polished and then prepared using the same process used for the quartz specimen. It was heated in the vacuum oven using a ceramic heater. Figure 9 represents a horizontal displacement pattern at 765°C, which was the highest recorded temperature for this specimen before the heater burned. The fringes are well defined. The small deviations from straightness are due to the nonhomogeneous nature of composite materials.

Figure 9.   Moiré pattern representing the horizontal displacement field due to thermal deformation of titanium matrix composite reinforced with silicon-carbide fibers specimen at 765°C.

At temperatures above about 700°C the thermal radiation from the specimen becomes a problem. It decreases the contrast of the interference patterns. Fortunately the thermal radiation is totally incoherent. It can be reduced to a negligible level by spatial filtering and narrow band filtering. Both methods were used in the experiments performed with quartz and metal matrix specimens. The patterns were photographed through a small aperture and a narrow band interference filter tuned to the 632nm wavelength of the HeNe laser.

Summary

It has been demonstrated that moiré interferometry can be applied at high temperatures without sacrificing its sensitivity and resolution. Routine measurements can be performed at 200°C using standard, replicated specimen gratings and an interferometer with extended working distance. If proper precautions are taken to minimize the convection currents and thermal distortions to the optics, this temperature range can be further extended to about 300°C with higher temperature replication media such as silicone rubbers. At higher temperatures, in the 300°C to 1000°C range, it is necessary to produce high temperature gratings using photoresist process as described above. In principle measurements can be performed at even higher temperatures. In this case gold used in the gratings described above must be replaced with metals like platinum or tungsten or eliminated. The research discussed herein shows that although the implementation of moiré interferometry in a vacuum oven is a little more expensive and requires different methods of heating the specimens, it fully eliminates errors caused by thermal currents. The application of an achromatic interferometer provides an automatic compensation for the change of the wavelength of light in a vacuum, so no major tuning is necessary after evacuating the oven.

Acknowledgments

The presented research was sponsored by Du Pont Company, the Virginia Center for Innovative Technologies and National Science Foundation. The experiments were performed together with Dr. Joosik Lee, Dr. Jau-Je Wu and Dr. Shih-Yung Lin. The author would also like to express appreciation to his wife Victoria C. Czarnek for help in editing this paper.

Literature Cited

(1)   J. Guild, "The Interference Systems of Crossed Diffraction Gratings; Theory of Moiré Fringes," Oxford at the Clardendon Press, (1956).

(2)   R. Czarnek, "Moiré Interferometry," chapter 11.2 of *Structural Testing: A Monograph for the Society for Experimental Mechanics*, edited by R. T. Reese and W. A. Kawahara, pp. 183-202, 1993.

(3)   R. Czarnek, "New Methods in Moiré Interferometry," Ph.D. Dissertation, Virginia Polytechnic Institute and State University, 1984.

(4)   R. Czarnek, "High-Sensitivity Moiré Interferometry with a Compact Achromatic Interferometer," *Optics and Lasers in Engineering*, Vol. 13, No. 2, pp. 99-115 (1990).

(5)   Robert Czarnek, Jau-Je Wu, Shih-Yung Lin and Joosik Lee, "High-Temperature, High-Sensitivity Moiré Interferometry," *Experimental Techniques*, Vol. 17, No. 3, pp. 29-33 (1993)

(6)   Jau-Je Wu, "Moire Interferometry at High Temperatures," Ph.D. Dissertation, Virginia Polytechnic Institute and State University, 1992.

(7)   Robert Czarnek, Joosik Lee and Tom Rantis, "Moiré interferometry with enhanced resolution," *Experimental Techniques*, Vol. 14, No. 4, pp. 24-28, July/August 1990.

(8)   J. Morton and R. Czarnek, "Dynamic Response of Advanced Material Systems: A Photomechanics Approach," Quarterly Report, DARPA/ONR project number N0014-88-K-0741, June-1989.

(9)   A. Kearney and C. Forno, "High Temperature Resistant Gratings For Moiré Interferometry," *Experimental Techniques*, Vol. 17, No. 6, pp. 9-12, November/December 1993.

RECEIVED June 9, 1995

Chapter 5

Positron Annihilation Lifetime Spectroscopy To Probe Free-Volume Effects in High-Temperature Polymers and Composites

A. J. Hill

Faculty of Engineering, Monash University, Clayton, Victoria 3168, Australia

Positron annihilation lifetime spectroscopy (PALS) as a probe of free volume in polymers and composites is reviewed. The effect of temperature on the free volume distribution and studies of the effects of thermal history on room temperature free volume and related mechanical, physical and diffusional properties are presented. The free volume related phenomena discussed include the glass transition, physical aging, gas permeability, and crosslink density.

Positron annihilation lifetime spectroscopy (PALS) provides an atomic scale probe of the free volume in polymers and polymeric composites. The free volume model applied to PALS [1] interprets the lifetime of orthoPositronium (oPs) localized in inter- and intra-molecular spaces as a measure of the size of those spaces. Typical oPs lifetimes in polymers range from 1 nanosecond (ns) to 3 ns corresponding to free volume cavity diameters ranging from 0.3 nm to 0.7 nm. These distances compare well with nonbonded interatomic distances in polymers and the root mean square end-to-end distance of an average polymer repeat unit.

Numerous properties of importance to the performance of polymers and composites at high temperatures are related to the polymeric free volume. Perhaps the most commonly quoted polymer and composite properties of interest at high temperatures are the mechanical and transport properties which are affected by thermal history [2,3]. High temperature effects on polymer and composite properties can occur during processing, in service, or due to combined effects of processing and in-service thermal history. These high temperature effects can be beneficial (e.g. post cure) or deleterious (physical aging) to the particular property of interest. For example, physical aging, which is the structural relaxation that occurs in glassy polymers, causes changes in mechanical properties (decrease in ductility, increase in yield strength) and diffusional properties which have been linked to a decrease in free volume [4,5].

Free volume concepts were developed in order to model the relationship between viscosity and volume in polymer melts. Polymer behavior in the melt can be extended to the glass transition region [6,7] and below T_g [8-10] by free volume theories that can

0097–6156/95/0603–0063$12.00/0

accurately model the effects of temperature, pressure and stress on the physical state of the polymer. As most engineering polymers have glass or melt transitions in the range 90°C to 180°C, operating temperatures above 50°C can lead to either stress relief, physical aging, or crystallization, each of which can alter the physical, mechanical, and diffusional properties of the polymer as well as cause dimensional changes. Models of physical aging [4,5], yield behavior [11], and permeability [12,13] are based on free volume, hence the PALS technique aids the development of an understanding, on the molecular level, of these and other free volume related phenomena.

Two aspects of PALS make the technique particularly useful for polymer and composite characterization. Firstly, PALS provides a direct probe of the free volume sites which are used to model the phenomena mentioned previously as well as the glass transition [14], blend miscibility [15], plasticization [16], stress relaxation [14], and diluent diffusion [17]. Secondly, PALS is a nondestructive characterization technique with the potential to monitor the state of the polymer or composite during processing or while in service. This chapter will discuss the PALS technique as a free volume probe and review some of the literature of PALS applied to polymers and composites.

The PALS Technique

Positrons are positively charged electrons that are emitted by some unstable isotopes during decay. Positron emitting isotopes are available commercially and the isotope most commonly used for PALS is ^{22}Na. This isotope emits a positron and a 1.28 MeV photon almost simultaneously through its nuclear decay to ^{22}Ne. In order to use positrons to study materials, the material of interest is placed next to the ^{22}Na positron source. The positrons are emitted with a range of energies (maximum energy of 540 keV) and penetrate a few hundred microns in polymeric materials. Each positron eventually annihilates with an electron in the material. PALS measures the distribution of positron lifetimes which is related to the electron density of the material at the positron-electron annihilation site. The positron is repelled by nuclei and hence prefers sites of lower than average electron density such as vacancy-type defects in metals or free volume cavities in polymers.

In order to determine the positron lifetime, a start and stop signal are used to mark the positron birth (emission from ^{22}Na) and death (annihilation with an electron) respectively. The 1.28 MeV photon that is emitted almost simultaneously with the positron is used to mark the positron birth. During annihilation, the masses of the positron and electron are converted into energy, typically through the emission of two 0.511 MeV photons. One of the 0.511 MeV photons is used to mark the positron death. The lifetime of the positron is dependent on its state and the local electron density at the annihilation site. In polymers, the positron can remain in its free state, become trapped at a site of lower than average electron density, combine with molecular species in the solid, or extract an electron from the surrounding material to form a semi-stable bound state known as positronium (Ps). Positronium can exist in two ground states: paraPositronium (pPs) in which the positron and electron have antiparallel spins, and orthoPositronium (oPs) in which the positron and electron have parallel spins.

The distribution of positron lifetimes is collected as a timing histogram of approximately one million sequential annihilation events in a material of interest. The positron decay curve (which contains the positron lifetimes and their relative intensities) is convoluted with the instrumental resolution function and contributions from source and background components. Computer programs are available to extract the positron lifetimes and relative intensities for the distinguishable positron states [18-20]. In PALS studies of polymers the decay curve is usually fitted with three lifetimes. The three lifetimes and their related annihilation mechanisms are commonly quoted as:

τ_1 0.1 - 0.3 ns   pPs self-annihilation and pPs-molecular species,

τ_2 0.3 - 0.5 ns   free positrons, positrons trapped in free volume, and
                                    positron-molecular species,

τ_3 0.5 - 4 ns   oPs pickoff annihilation in free volume.

OrthoPositronium pickoff annihilation is a quenching process during which the positron in oPs (localized in free volume cavities) annihilates with an electron, of opposite spin, from the surrounding cavity wall.

The PALS equipment is often a fast-fast coincidence system comprised of hardware supplied by EG&G Ortec that is either thermally or digitally stabilized [21]. The ^{22}Na source is typically available as aqueous ^{22}NaCl which is evaporated, drop by drop, on a thin (< 3 mg cm^{-2}) metal foil creating a spot source of 2 mm diameter and activity of 0.5 to 2 MBq. For sorption studies an impermeable thin film material is used in place of the metal foil. A second foil is placed on the source foil, and the foils are sealed (by crimping or adhesive). The foil is sandwiched between two identical pieces of the sample material such that all positrons emitting from the source enter the sample material. Depending on sample thickness (at least 0.5 mm) and source strength, it takes approximately 15 minutes to 2 hours to collect one spectrum for analysis. Typically several spectra are collected such that population standard deviations can be reported on the lifetime and intensity parameters. The population standard deviations usually quoted for the oPs pickoff components in polymer studies are +/- 0.03 ns for the lifetime, τ_3, and +/- 0.3 % for the intensity, I_3. The fitting programs for data analysis can model the lifetime spectrum as a finite sum of decaying exponentials [18] or can perform a Laplace inversion of the decay curve and extract the continuous lifetime distribution [22]. The number of counts needed for statistical accuracy of the Laplace inversion is much greater than that needed for acceptable results from the exponential fitting routines resulting in measurement times an order of magnitude longer in order to extract continuous lifetime distributions [23].

Radiation safety is a primary concern when dealing with radionuclides and ionizing radiation. Although ^{22}Na is a high level isotope due to the 1.28 MeV photon emission, the source strengths used in PALS are relatively low, and hence laboratories should meet a medium level standard by installing a barrier and an ionizing radiation monitor. The barrier should restrict laboratory access to radiation workers only, and the monitor should be available to measure radiation levels. Some samples may leach ^{22}Na from the source and become contaminated. These samples must be bagged, labeled and properly disposed. Samples and sources should be handled using disposable gloves. Sources not in use should be stored in lead containers; lead shielding can be used around the source-sample sandwich avoiding obstruction of the detector heads. Individual radiation monitors (film badges or thermoluminescent devices) should be worn by personnel working in a PALS laboratory.

PALS as a Free Volume Probe

Usually the third lifetime component (τ_3, I_3) is associated with free volume related phenomena in polymers where τ_3 is related to the mean radius of the free volume sites and I_3 is related to the concentration of free volume sites. There are, however, several free volume models for positron annihilation in polymers, some of which combine the second and third lifetime components in order to calculate a free volume size or fraction [1,24-29]. This section provides a generally accepted literature view of the free volume model applied to PALS based on the original work by Brandt et al. [1]. The Brandt free volume model was proposed in order to explain the observed increase in oPs lifetime with increasing temperature in molecular solids [1]. From rate kinetics considerations one might expect a decrease in lifetime (increase in annihilation rate) with increasing temperature; however, Brandt et al. [1] proposed a quantum mechanical explanation for the observed increase in oPs lifetime. According to the model, oPs localizes in free volume cavities in the polymer "lattice." The oPs lifetime is related to the probability of overlap of the oPs wave function with the wave function of an electron associated with the surrounding cavity wall. As the size of the free volume site increases, the local electron density decreases and the oPs lives longer resulting in an increase in τ_3. The intensity of oPs annihilations, I_3, increases with an increase in population of free volume sites due to a higher oPs formation and trapping rate. Thus PALS can yield information on the relative size and relative concentration of free volume sites; however, care must be taken in the interpretation of PALS results for molecular solids. For example, contact time with the ^{22}Na source can create competitive processes (electron or positron scavengers, etc.) in some polymers which inhibit Ps formation thus reducing I_3. The chemical properties of some functional groups also can affect Ps formation due to their ability to scavenge electrons or their electron or positron affinity [30]. In most cases Ps formation is not completely inhibited, and τ_3 remains unaffected. Hence τ_3 reflections of free volume cavity radius are less subject to the non-free volume related effects of radiation chemistry than I_3 reflections of free volume concentration.

Models of lifetime related to a mean free volume cavity radius, R, have been developed [25,31], and the free volume size is calculated as $V_f = 4\pi R^3/3$. A Laplace inversion of the positron lifetime decay curve allows the continuous lifetime distribution to be obtained in place of the average lifetimes [22,23]. In the context of the free volume model, the free volume cavity size distribution is obtained in place of the mean free volume cavity size. The distribution of free volume sizes is important to free volume models of polymer behavior because only a part of the total free volume is available for molecular or segmental motion. Several recent PALS studies have been performed to investigate the effect of temperature [32,33], hydration [34], and pressure

[35,36] on the free volume size distribution. Other free volume probes such as fluorescent, electrochromic, and photochromic probes are, in general, larger than the intrinsic free volume cavities of polymers and measure different free volume distributions than the PALS probe [33,37]. This chapter will illustrate the applicability of the PALS results to free volume related properties of interest for high temperature polymeric materials.

PALS Applied to Polymers and Composites

Molecular Relaxations and Free Volume Distributions. The classical free volume theories for liquids by Doolittle [6] and Fox and Flory [7] were developed for thermodynamic equilibrium and have not been successfully extended to temperatures far below T_g. An equation of state for glassy polymers, developed by Simha and Somcynsky [8] and subsequently modified [9], has successfully modeled the effects of temperature, pressure and stress on the free volume in the glassy state [10,38,39]. Comparison of the Simha and Somcynsky free volume parameter to the free volume parameters measured by PALS shows reasonable agreement [33,40]. PALS free volume measurements as a function of temperature have been used to study T_g and sub-T_g (T_β, T_γ) mobility - free volume relationships in numerous polymers such as epoxy [41], polycarbonate [42], polyethylene terephthalate [43], polytetrafluoroethylene [44] polypropylene [45], polyethylene [46], and polystyrene [46,47,48]. The work of Kasbekar [48], shown in Figures 1 and 2, illustrates the dependence of the oPs pickoff annihilation parameters, τ_3 and I_3, on temperature. The glass transition is indicated as a change in slope of τ_3 versus temperature similar to specific volume - temperature behavior. The I_3 parameter, shown in Figure 2, also is sensitive to molecular motion and indicates a change in the relative number of free volume cavities rather than a change in their mean size. The glass transition indicated by PALS is usually evident at a lower temperature than that found by differential scanning calorimetry or dynamic mechanical analysis. It is postulated that the oPs free volume probe detects the changes in electron density due to molecular motion that occur at temperatures lower than the cooperative motion of chains measured by other techniques. The annihilation rate of oPs by pickoff in free volume cavities is approximately 10^{10} Hz, hence molecular motions occurring at a lower frequency occupy dynamic free volume that is accessible to oPs. As the frequency of motion changes with temperature, these sites can become inaccessible to oPs, and this change in dynamic free volume allows the detection of sub-T_g transitions by PALS.

As polymers are cooled through T_g, the inability of molecular relaxations to occur on the time scale of cooling creates a nonequilibrium state, the glassy state. In the glassy state, thermodynamic properties change at constant temperature and pressure as a function of time. The gradual approach of polymer properties toward an equilibrium below T_g is termed physical aging. Physical aging can be accelerated by annealing glassy polymers at temperatures close to but below T_g. The extent of physical aging is

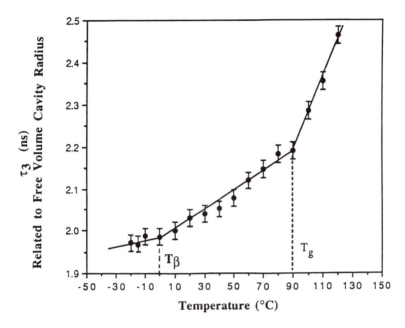

Figure 1. The oPs pickoff lifetime, τ_3, as a function of temperature for atactic polystyrene. Adapted from ref. 48.

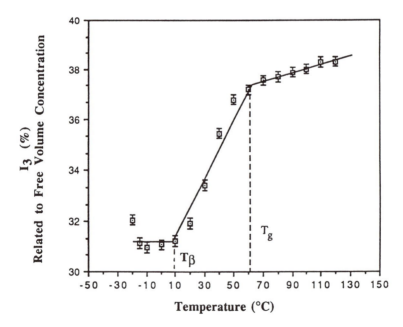

Figure 2. The oPs pickoff intensity, I_3, as a function of temperature for atactic polystyrene. Adapted from ref. 48.

a function of aging time and temperature and is reflected in physical property changes, such as an increase in density and T_g, and in mechanical property changes, such as an increase in tensile yield strength and a decrease in fracture toughness. As mentioned previously the extension of free volume theories below T_g has been used to link molecular mobility with physical and mechanical property changes due to physical aging [4,10]. PALS measurements of the free volume change due to physical aging and the relaxation kinetics of free volume in the glassy state have been linked to molecular mobility, the structural state, and hence physical and mechanical properties [49-52]. Sensitivity of the oPs pickoff component, (τ_3, I_3) to structural relaxation (physical aging) in the glassy state has been reported in epoxy [50], polycarbonate [53], polyvinylacetate [54], and polystyrene [49].

The distribution of hole sizes probed by PALS ranges from the macroscopic pore size range in porous epoxy resins (and other high surface area materials) [55] to the atomic scale free volume cavities discussed thus far. Surface area studies of porous resins will not be discussed in this review; however, these measurements are useful for polymer and composite characterization in relation to moisture uptake [20]. The continuous distribution of positron lifetimes (free volume radii) can be calculated from PALS data using the program CONTIN(PALS2) [56]. Studies of the free volume distribution for polymers as a function of temperature [32,33], pressure [35,36], and hydration [34] have been reported. For example the temperature dependence of free volume in polyoxymethylene has been reported by Kristiak et al [32]. Figure 3 illustrates the change in the free volume distribution probed by PALS as temperature is varied. The lifetime τ_3 has been converted to a free volume cavity radius R using the model of Nakanishi et al. [25]. Figure 3 shows that the free volume cavity radii decrease as temperature is lowered and their distribution narrows. Deng and Jean [36] have measured the free volume distribution in an epoxy glass as a function of pressure. The second and third lifetime components are shown (Figures 4 and 5) as both can be related to free volume [26,31,57]. Increasing the pressure decreases the mean free volume cavity radii and narrows the distribution of free volume sizes. Polymers and composites usually are subject to thermal cycling and/or moist environments while under load (tensile or compressive), and knowledge of the free volume distribution as a function of temperature and stress can help predict service properties such as permeability, sorption, and stress corrosion cracking resistance.

Gas Permeability. The distribution of free volume sizes is important to transport and sorptive properties, and the PALS free volume probe can be used to measure a particular fraction of the total static and dynamic free volume [33,58]. Water absorption in polymers and composites has been studied by PALS to determine the sensitivity of the oPs pickoff component (τ_3, I_3) to equilibrium moisture uptake [59-64], moisture uptake rate [65], the states of water, whether bound or free [59,66], and the effect of thermal history on moisture uptake [62]. These studies have been performed on polyimides [62], epoxies [59-65], nylons [62], and polyvinylalcohol [66].

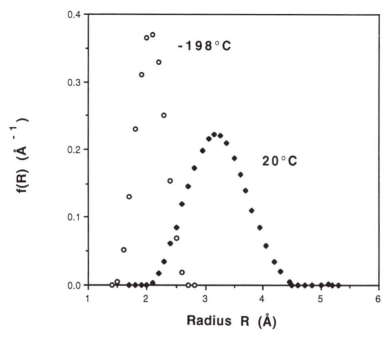

Figure 3. The free volume cavity radius distribution function f(R) as a function of cavity radius $R(\tau_3)$ at two temperatures for semicrystalline polyoxymethylene. Adapted from ref. 32.

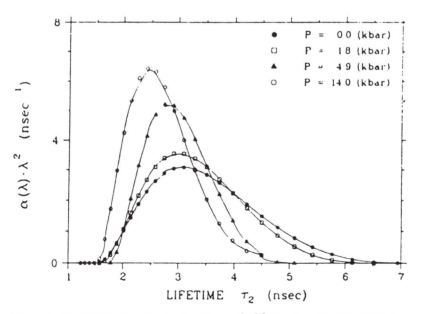

Figure 4. The distribution density function, $\alpha(\lambda) \cdot \lambda^2$, for the oPs pickoff lifetime, τ_2, as a function of τ_2. Reproduced with permission from ref. 36. Copyright 1993 American Chemical Society.

Experimental evidence has been compiled suggesting that the free volume probed by PALS is related to permeability [37,67] physical aging [49-54], fatigue crazing [68,69], plasticization [70,71] water absorption [59-66], miscibility of polymer blends [72,73], interfacial phenomena [74-76], phase separation [77], degree of crystallinity [31,58], the glass transition [41-48], and sub-T_g transitions [41,46,48].   Sorption and permeability studies provide a convenient test of the postulated relationships between free volume and diffusional properties [37,67].

The free volume theory of Cohen and Turnbull [13] as discussed by Ratner [77] and Kobayashi et al. [67] models diffusion in polymers by the following equation:

$$D = A\exp[-\gamma V^*/V_f] \tag{1}$$

where D is the diffusivity, A and γ are constants related to the diffusing molecule, V^* is the critical void volume necessary for diffusive displacements, and V_f is the average free volume per molecule.   As V_f increases, the probability of free volume cavities existing with volumes V^* or greater increases, and diffusion is facilitated.   Using the free volume model of Brandt et al. [1], the oPs pickoff lifetime, τ_3, is related to free volume cavity radius; the volume of the cavity is then related to the quantity τ_3^3.

A recent study of gas diffusivity used PALS to probe the free volume - diffusivity relationship in several polymers [67] and provides a good example for illustrative purposes.   In their work on gas diffusivity, Kobayashi et al. [67] converted τ_3 to a radius dimension using the model of Nakanishi et al. [25] and calculated cavity volumes showing that lnD is proportional to $1/V_f$ for numerous polymers and various gases as predicted by Equation 1.   For the purpose of the present review, the lifetime data of Kobayashi et al. [67] are presented in raw form in order to show the τ_3 and $1/\tau_3^3$ relationship to diffusivity of molecular oxygen in various polymers at room temperature. The data are presented in Figures 6 and 7.   The abbreviations are defined as follows: PES=polyethersulfone, PSF=polysulfone, PC=polycarbonate, PS= polystyrene, PETU=polyetherurethane, PMP=polymethylpentene, PDMS=polydimethyl -siloxane.   Some of these polymers have T_g's below room temperature hence the relationship in Equation 1 appears to hold for the glassy and rubbery state.   In addition, some of the polymers are partially crystalline.   The effect of increased crystallinity reducing diffusivity is well known [79].   Most reports of PALS studies of crystallinity show constant τ_3 and decreasing I_3 with increasing crystallinity leading to the postulate that most oPs pickoff annihilations take place in the amorphous regions or at amorphous/crystalline interfaces.   Molecular diffusion is expected to take place in the amorphous regions and at interfaces, hence the combined sensitivity of τ_3 to free volume size and I_3 to degree of crystallinity and interfacial regions makes the PALS technique a unique probe of that part of the free volume which affects transport and sorptive properties.   The remarkable correlation of $1/\tau_3^3$ with diffusivity shown in Figure 7 (data of Kobayashi et al. [67]) supports the claim that the free volume distribution probed by PALS is important to transport properties.

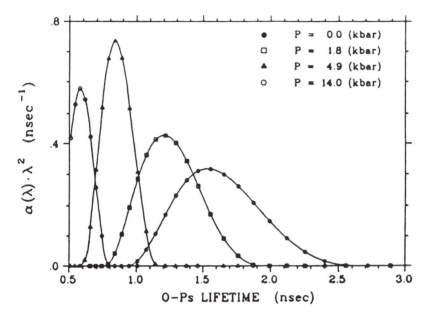

Figure 5. The distribution density function, $\alpha(\lambda)\cdot\lambda^2$, for the positron lifetime, τ_3, as a function of τ_3. Reproduced with permission from ref. 36. Copyright 1993 American Chemical Society.

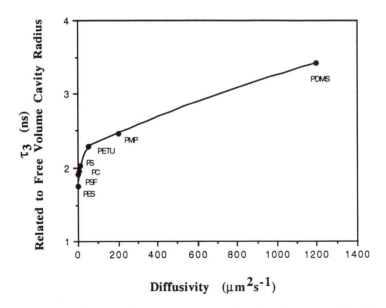

Figure 6. The oPs pickoff lifetime, τ_3, as a function of oxygen diffusivity in several polymers (defined in the text) at room temperature. Adapted from ref. 67.

Crosslink Density. Numerous studies of PALS applied to thermosets (mainly epoxy systems) have been reported. These studies have examined the effect on PALS free volume of moisture uptake [59-65], pressure [36,57], temperature [41,80], variable fiber fraction [81], chemistry [59,64,82,83], molecular weight between crosslinks [59,78,82,83], physical aging [50], plasticization [71], degree of cure [81], and temperature of cure [64,83]. The effect of changing molecular weight between crosslinks, M_c, on the PALS free volume has been investigated by Singh et al. [59]. The free volume model of Singh and Eftekhari [26] computes the effective or average free volume cavity size using four of the PALS parameters, namely τ_2, I_2, τ_3, and I_3. The model has been used successfully to predict molecular weights of linear polymers [26,84] and the molecular weight between crosslinks in several epoxies [59]. The effective free volume cavity size is calculated from the equation:

$$V_f = \frac{V_{f_2} I_2 + V_{f_3} I_3}{I_2 + I_3} \qquad (2)$$

where the intensities, I_2 and I_3, are from the experimental data, and V_{f_2} and V_{f_3} are based on radii derived from the semi-empirical model of Nakanishi et al. [25] relating positron lifetime to a spherical free volume cavity radius. Using both τ_2 and τ_3 allows modeling of a bimodal free volume distribution with τ_2 representing the smaller cavities. It was shown previously (Figures 4 and 5) that the τ_2 and τ_3 components are dependent on pressure; both lifetimes decrease with increasing pressure in a way that indicates annihilations taking place in free volume cavities [57]. In the study of epoxies by Singh et al. [59] the molecular weight between crosslinks, M_c, was calculated from the known ratios of the diprimary and disecondary amines in the stoichiometric epoxy resins. The calculations were based on complete chemical reaction and were confirmed by swelling experiments. The Singh-Eftekhari model predicts the relationship

$$V_f = A M_c^B \qquad (3)$$

where A and B are structural constants. A comparison of the fit provided by Equation 3 and the PALS data in the form of Equation 2 is shown in Figure 8. PALS characterization of the cured system is useful for quality control and property prediction, and as mentioned previously, the effects of chemistry and thermal history on the network structure, and hence properties, are areas of active PALS research [41,59,64,82,83]. These recent studies have reported the influence of chain flexibility between crosslinks on T_g and packing, the effect of cure temperature on T_g and free volume concentration, and the effect of crosslink density on T_g and sub-T_g transitions as measured by PALS.

Network Formation. Another area of importance in thermoset resins and composites is the characterization of network formation and monitoring of cure as it proceeds.

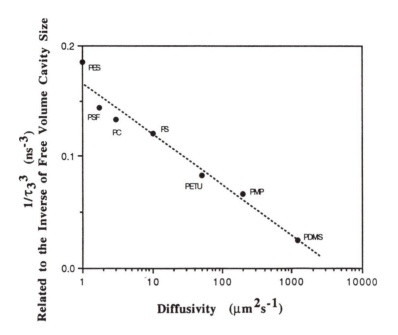

Figure 7. The inverse of the cube of the oPs pickoff lifetime, $1/\tau_3^3$, (related to $1/V_f$) as a function of the natural logarithm of oxygen diffusivity showing reasonable agreement with Equation 1. Polymer abbreviations are defined in the text. Adapted from ref. 67.

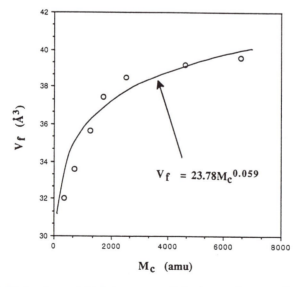

$$V_f = 23.78 M_c^{0.059}$$

Figure 8. Molecular weight between crosslinks in epoxies as a function of the effective free volume cavity size calculated from the free volume theory of Singh and Eftekhari. Adapted from ref. 59.

Much effort has been devoted to the development of real time cure sensors based on dielectrics, ultrasonics, and fiber optics. The collection times of 15 minutes to 2 hours for each PALS spectrum precludes real time monitoring for industrial applications; however, laboratory research at lower temperatures (to slow reaction kinetics) can contribute to the understanding of the role of free volume in network development. The work of Suzuki et al. [80] on epoxies illustrates the sensitivity of the oPs pickoff components (τ_3 and I_3) to the network development during a cure cycle. The results are shown in Figures 9 and 10. Curing takes place at 70°C over a 30 hr period followed by postcures at 100°C and 150°C. With the initial temperature increase, τ_3 and I_3 increase due to thermal expansion. As polymerization begins, there is a decrease in τ_3 and a slope change in I_3 indicative of the consumption of monomer and the change in packing. Gelation is marked by a leveling off of both the τ_3 and I_3 parameters as the 3-D network is formed. In this study, further cure and postcure are evident only from the PALS parameter response to temperature change. On final cooling of the glass, the τ_3 value returns to that characteristic of gelation; however, I_3 indicates retained free volume concentration (excess free volume) which is time dependent. The time dependent I_3 response to temperature or pressure change has been characterized in other glassy polymers and related to structural relaxation [31,53]. The relaxation kinetics are indicative of the state of the glass and have been used to characterize physical aging as discussed in a previous section of this review. The lifetime τ_3 and intensity I_3 results show that the PALS parameters related to free volume can detect the commencement of polymerization, propagation of the 3-D network, termination of polymerization, and are sensitive to the cure temperature and thermal history.

Conclusions and Future Work

This review has presented and discussed a number of examples of PALS studies of polymers and composites. PALS gives a measure of the positron lifetime which is related to the electron density at the annihilation site. Changes in chemistry, temperature, pressure or some other variable which affect the electron density at the oPs localization (annihilation) site are reflected in the oPs pickoff lifetime. The oPs localizes in regions of reduced electron density such as free volume and interfacial sites in polymers and composites. Hence the variations in the lifetime, τ_3, and intensity, I_3, of the oPs pickoff component have been related to free volume behavior for the examples discussed in this review. The sensitivity of the PALS probe to the free volume that affects physical, diffusional, and mechanical properties indicates that the PALS technique provides a useful characterization tool. The measurement of free volume distributions as functions of temperature, pressure, hydration, or stress can provide information to help understand the mechanisms of polymer and composite degradation under load in hot-wet environments, leading to the design of better polymers, fillers, and interfaces.

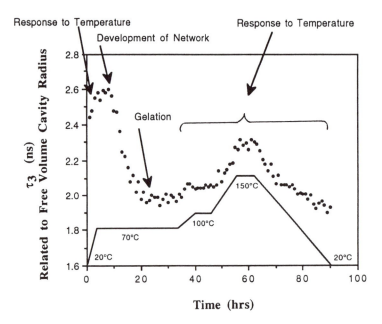

Figure 9. The oPs pickoff lifetime, τ_3, as a function of time during the curing cycle for an epoxy. Adapted from ref. 80.

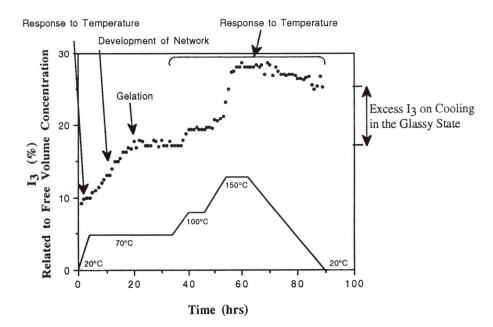

Figure 10. The oPs pickoff intensity, I_3, as a function of time during the curing cycle for an epoxy. Adapted from ref. 80.

PALS studies of moisture uptake in thermosets have indicated that larger free volume cavities lead to higher absorption rates [65]. The distribution of free volume sizes is obviously important to diffusional properties, and as noted by Jeffrey and Pethrick [64], the presence of submicron and micron size pores contributes significantly to the moisture uptake. The oPs pickoff component can be used to measure pores as well as atomic-scale free volume cavities; hence the technique is useful for the prediction of hot-wet behavior. Also important to composites is the interfacial free volume. Relatively few PALS studies of composites have been reported [63,81,85]; however, PALS sensitivity to interfacial effects in phase separated homopolymers and polymer blends has been reported [74-77]. Systematic studies of composites with varying interfacial chemistry and thermal history could help in the identification of annihilations at interfaces and correlation of free volume or chemistry related interfacial effects with composite properties. The ability to follow cure reactions as functions of time and temperature is important to process control. In epoxy cure, it is the final etherification reactions which are inaccessible to DSC characterization of cure but followed rather well by near infrared and dielectric loss spectroscopies. PALS studies in conjunction with near infrared and dielectric loss could shed light on the effect of these reactions on the network structure and free volume related properties.

The depth of penetration of positrons emitted from ^{22}Na is approximately a few hundred microns in polymers. The development of slow and/or monoenergetic positron beams [86] allows depth profiling of materials and should provide useful results for polymers and composites including network inhomogeneities due to cure history, fiber placement, damage sites, weak interface, etc. The recent and forthcoming advances in the use of PALS to characterize polymers and composites should help scientists and engineers progress in their understanding of free volume and its effect on properties.

Acknowledgments

The author would like to thank J. Gonis, M. D. Zipper, R. M. Hodge, G. P. Simon, K. J. Heater, A. Siegle, H. Schneider and T. J. M. Bastow for reference help and useful discussions.

Literature Cited

1.   Brandt, W.; Berko, S.; Walker, W. *Phys. Rev.* **1960**, *120*, 1289.
2.   Tant, M. R.; Henderson, J. in *International Encylcopaedia of Composites*, Vol. 6, Lee, S. M.; ed. VCH Publishers: New York, 1990.
3.   Hollaway, E., Ed. *Handbook of Polymer Composites for Engineers*, Woodhead Publ. Ltd.: England, 1994.
4.   Struik, L. C. E. *Ann. N. Y. Acad. Sci.* **1976**, *279*, 78.
5.   Struik, L. C. E. *Physical Aging in Amorphous Polymers and Other Materials*, Elsevier: Amsterdam, 1978.
6.   Doolittle, A. K. *J. Appl. Phys.* **1951**, *22*, 471.
7.   Fox, T. G.; Flory, P. J. *J Amer. Chem. Soc.* **1948**, *70*, 2384.
8.   Simha, R.; Somcynsky, T. *Macromolecules* **1969**, *2*, 342.

9.   Neis, E.; Stroeks, A. *Macromolecules* **1990**, *23*, 4088.
10. Curro. J. G.; Lagasse, R. R., Simha, R. *J. Appl. Phys.* **1981**, *52*, 5892.
11. Glasstone, S.; Laidler, K. J.; Eyring, H. *The Theory of Rate Processes*, McGraw-Hill: New York, 1941.
12. Maeda, Y.; Paul, D. R. *J. Polym. Sci. B* **1987**, *25*, 1005.
13. Cohen, M. H.; Turnbull, D. *J. Chem. Phys.* **1959**, *31*, 1164.
14. Ferry, J. D. *Viscoelastic Properties of Polymers*, John Wiley and Sons: New York, 1980.
15. Patterson, D.; Robard, A. *Macromolecules* **1978**, *11*, 690.
16. Doolittle, A. K. *The Technology of Solvents and Plasticisers*, Wiley: New York, 1954.
17. Vrentas, J. S.; Duda, J. L.; Ling, H.-C. *J. Polym. Sci. B* **1985**, *23*, 275.
18. Puff, W. *Comput. Phys. Commun.* **1983**, *30*, 359.
19. Seigel, R. W. *Ann. Rev. Mater. Sci.* **1980**, *10*, 393.
20. Schrader, D. M; Jean, Y. C, Eds. *Positron and Positronium Chemistry: Studies in Physical and Theoretical Chemistry*, Elsevier: New York, 1988.
21. Paulus, T. J. *Optimization of a State-of-the-Art Positron Measurement Apparatus,* EG&G Ortec, Oakridge, TN, 1988.
22. Gregory, R. B. *J. Appl. Phys.* **1991**, *70*, 4665.
23. Jean, Y. C.; Dai, G. H. *Nucl. Instrum, Methods Phys. Res. B* **1993**, *B79*, 356.
24. Tao, S. J. *J. Chem. Phys.* **1972**, *56*, 5499.
25. Nakanishi, H.; Wang, S. J.; Jean, Y. C. *International Symposium on Positron Annihilation Studies of Fluids*, Sharma, S. C. Ed., World Scientific: Singapore, 1987, 292.
26. Singh, J. J.; Eftekhari, A. *Nucl. Instrum. Methods Phys. Res. B* **1992**, *B63*, 477.
27. Eldrup, M.; Lightbody, D.; Sherwood, J. N. *Chem. Phys.* **1981**, *63*, 51.
28. Thosar, B. V.; Kulkarni, V. G.; Lagu, R. G.; Chandra, G. *Phys. Lett.* **1969**, *28A*, 760.
29. Gol'danskii, V. I.; Onishchuk, V. A; Shantorovich, V. P.; Volkov, V. V; Yampol'skii, Y. P. *Khim. Fiz.* **1988**, *7*, 616.
30. Mogensen, O. E. *J. de Physique IV*, **1993**, *3*, 1.
31. Jean, Y. C. *Microchem. J.* **1990**, *42*, 72.
32. Kristiak, J.; Kristiakova, K.; Sausa, O.; Bandzuch, P.; Bartos, J. *J. de Physique IV*, **1993**, *3*, 265.
33. Lui, J.; Deng, Q.; Jean, Y. C. *Macromolecules* **1993**, *26*, 7149.
34. Gregory, R. B; Chai, K.-J. *J. de Physique IV*, **1993**, *3*, 305.
35. Jean, Y. C.; Deng, Q. *J. Polym. Sci. B* **1992**, *30*, 1359.
36. Deng, Q.; Jean, Y. C. *Macromolecules* **1993**, *26*, 30.
37. Yampol'skii, Y. P.; Shantorovich, V. P.; Chernyakovskii, F. P.; Kornilov, A. I.; Plate, N. A. *J. Appl. Polym. Sci.* **1993**, *47*, 85.
38. Robertson, R. E.; Simha, R.; Curro, J. G. *Macromolecules* **1984**, *17*, 911.
39. McKinney, J. E.; Simha, R. *Macromolecules* **1976**, *9*, 430.
40. Vleeshouwers, S.; Kluin, J.-E.; McGervey, J. D.; Jamieson, A. M.; Simha, R. *J. Polym. Sci. B* **1992**, *30*, 1429.
41. Jean, Y. C.; Sandreczki, T. C.; Ames, D. P. *J. Polym. Sci. B* **1986**, *24*, 1247.
42. Kluin, J.-E.; Yu, Z.; Vleeshouwers, S.; McGervey, J. D; Jamieson, A. M.; Simha, R. *Macromolecules* **1992**, *25*, 5089.
43. Tseng, P.; Chang, S.; Chuang, S. *Positron Annihilation (Proc. of 6th International Conference on Positron Annihilation)* Coleman, P. G.; Sharma, S. C.; Diana, L. M. Eds., North Holland: New York, 1982, 730.
44. Kindl, P.; Sormann, H.; Puff, W. *Positron Annihilation (Proc. of 6th International Conference on Positron Annihilation)* Coleman, P. G.; Sharma, S. C.; Diana, L. M. Eds., North Holland: New York, 1982, 685.
45. Lind, J. H.; Jones, P. L.; Pearsall, G. W. *J. Polym. Sci. A* **1986**, *24*, 3033.

46. Varisov, A. Z.; Kuznetsov, Y. N.; Prokop'ev, E. P.; Filip'ev, A. I. *Russian Chem. Rev.* **1981**, *150*, 991.
47. Stevens, J. R.; Mao, A. C. *J. Appl. Phys.* **1970**, *41*, 4273.
48. Kasbekar, A. D. *M. Sc. Thesis*, Duke University, Durham, NC, 1987.
49. McGervey, J.; Panigrahi, N.; Simha, R.; Jamieson, A. *Positron Annihilation (Proc. of 7th International Conference on Positron Annihilation)* Jain, P. C.; Singru, R. M; Gopinathan, K. P.. Eds., World Scientific Publ. Co.: Singapore, 1985, 690.
50. Sandreczki, T. C; Nakanishi, H.; Jean, Y. C. *International Symposium on Positron Annihilation Studies of Fluids*, Sharma, S. C. Ed., World Scientific: Singapore, 1987, 200.
51. Hill, A. J.; Katz, I. M.; Jones, P. L. *Polym. Sci. Engr.* **1990**, *30*, 762.
52. Heater, K. J.; Jones, P. L. *Mat. Res. Soc. Symp. Proc.* **1991**, *215*, 207.
53. Hill, A. J.; Heater, K. J.; Agrawal, C. M. *J. Polym. Sci. B* **1990**, *28*, 387.
54. Kobayashi, Y.; Zheng, W.; Meyer, E. F.; McGervey, J. D.; Jamieson, A. M.; Simha, R. *Macromolecules* **1989**, *22*, 2302.
55. Venkateswaran, K.; Cheng, K. L; Jean, Y. C. *J. Phys. Chem.* **1983**, *88*, 2465.
56. Gregory, R. B.; Shu, Y. *Nucl. Instrum. Methods Phys. Res. A* **1990**, *A290*, 172.
57. Deng, Q.; Sundar, C. S.; Jean, Y. *J. Phys. Chem.* **1992**, *96*, 492.
58. Simon, G. P.; Zipper, M. D.; Hill, A. J. *J. Appl. Polym. Sci.* **1994**, *52*, 1191.
59. Singh, J. J; Eftekhari, A.; Schultz, W. J.; St. Clair, T. L. *NASA Technical Memorandum 4390*, **1992**.
60. MacQueen, R. C.; Granata, R. D. *Mater. Sci. Forum* **1992**, *105-110*, 1649.
61. MacQueen, R. C.; Granata, R. D. *J. Polym. Sci. B* **1993**, *31*, 971.
62. Singh, J. J. Proc. *33rd Internatl. SAMPE Symp.*, 7-10 March 1988, SAMPE: Covina, CA, 407.
63. Singh, J. J.; Holt, W. H.; Mock, W. Jr. *NASA Technical Paper 1681*, **1980**.
64. Jeffrey, K.; Pethrick, R. A. *Eur. Polym. J.* **1993**, *30*, 153.
65. Suzuki, T.; Oki, Y.; Numajiri, M.; Miura, T.; Kondo, K.; Shiomi, Y.; Ito, Y. *J. Appl. Polym. Sci.* **1993**, *49*, 1921.
66. Hodge, R. M.; Bastow, T. J.; Edward, G. H.; Simon, G. P.; Hill, A. J. *Proc. 3rd Pacific Polym. Conf.* 13-17 Dec. 1993, Pacific Polym. Fed.: Gold Coast, Australia, 451.
67. Kobayashi, Y.; Haraya, K.; Hattori, S.; Sasuga, T. *Polymer* **1994**, *35*, 925.
68. Lui, L. B.; Lewis, J. C.; Yee, A. F.; Gidley, D. W. *8th Intl. Conf. on Deformation, Yield, and Fracture of Polymers*, 8-11 Apr. 1991, Cambridge UK, 29/1.
69. Lui, L. B.; Gidley, D. W.; Yee, A. F. *J. Polym. Sci. B* **1992**, *30*, 231.
70. Elwell, R. J.; Pethrick, R. A. *Eur. Polym. J.* **1990**, *26* 853.
71. Okada, T.; Nishijima, S.; Honda, Y.; Kobayashi, Y. *J. de Physique IV*, **1993**, *3*, 291.
72. Zipper, M. D.; Simon, G. P.; Cherry, P.; Hill, A. J. *J. Polym. Sci. B* **1994**, *32*, 1237.
73. Zipper, M. D.; Simon, G. P.; Tant, M. R.; Small, J. D.; Stack, G. M.; Hill, A. J. *Polym. Intl.* in press.
74. Zipper, M. D.; Simon, G. P.; Flaris, V.; Campbell, J. A.; Hill, A. J. *Mater. Sci. Forum*, in press.
75. Wang, C. L.; Wang, S. J; Zheng, W. G.; Qi, Z. N. *J. de Physique IV*, **1993**, *3*, 261.
76. Ravindrachary, V.; Sreepad, H. R.; Chandrashekara, A.; Ranganathaiah, C.; Gopal, S. *Phys. Rev. B* **1992**, *46*, 11471.
77. Chanyuan, Y.; Weizheng, X.; Qingchao, G.; Dexun, S. *Positron Annihilation (Proc. of 8th International Conference on Positron Annihilation)* 29 Aug. - 3 Sept., 1988, Gent, Belgium, World Scientific Publ. Co.: Singapore, 1989, 573.

78. Ratner, M. *Polymer Electrolyte Reviews*, MacCallum, J. R.; Vincent, C. A., Eds., Elsevier: London, 1987, 185.
79. Peterlin, A. *Interrelations between Processing Structure and Properties of Polymeric Materials*, Seferis, J. C.; Theocaris, P. S., Eds., Elsevier Science Publ. B. V.: Amsterdam, 1984, 585.
80. Suzuki, T.; Oki, Y.; Numajiri, M.; Miura, T.; Kondo, K.; Ito, Y. *Polymer* **1993**, *34*, 1361.
81. Dale, J. M; Hulett, L. E.; Roseel, T. M.; Fellers, J. F. *J. Appl. Polym. Sci.* **1987**, *33*, 3055.
82. Waber, J. T.; Wang, R.; Williams, J. *Positron Annihilation (Proc. of 9th International Conference on Positron Annihilation)* Sombathy, World Scientific Publ. Co.: Singapore, 1991, 1761.
83. Gonis, J.; Simon, G. P.; Hill, A. J. *Proc. 3rd Pacific Polym. Conf.* 13-17 Dec. 1993, Pacific Polym. Fed.: Gold Coast, Australia, 533.
84. Singh, J. J.; Eftekhari, A.; Hinkley, J. A.; St. Clair, T. L.; Jensen, B. J. *NASA Technical Memorandum 4326*, **1992**.
85. Tino, J.; Kristiak, J.; Hlouskova, Z.; Sausa, O. *Eur. Polym. J.* **1993**, *29*, 95.
86. *Positron Beams for Solids and Surfaces*; Schultz, P.; Massoumi, G. R.; Simpson, P. J., Eds.; American Institute of Physics: London, Ontario, 1990.

RECEIVED November 17, 1994

PROCESSING AND MODELING

Chapter 6

Some Models of Material Behavior During Injection Molding

M. R. Kamal, W. Frydrychowicz, and I. Ansari

Department of Chemical Engineering, McGill University,
Montreal H3A 2A7, Canada

Various models of polymer behavior during thermoplastic injection molding are considered. A computer simulation is described for injection mold filling and cooling in complex shaped cavities. Both fountain flow and coupling of crystallization kinetics within the equation of transport have a profound influence on the predictions of the thermo-mechanical history experienced by the material. A summary of some polymer crystallization kinetics models is included. Comparisons are made between the Avrami and Tobin models. Some practical relationships are outlined for evaluating the effects of induction time, pressure, and stress on crystallization kinetics. Direct and inverse models are outlined to estimate thermal stresses and the optimal temperature or pressure conditions during injection molding required to obtain a prescribed distribution of the thermal stresses.

The processing of polymer melts at high temperature involves a combination of simultaneous flow and heat transfer of the non-Newtonian, viscoelastic liquid accompanied with solidification of the partially crystalline material. In many cases, the flow system involves complex geometry and free surface flow. The thermo-mechanical history experienced by the polymer during the process produces a microstructure (orientation, crystallinty, morphology, thermal and flow stresses, etc.) which determines, to a large extent, the ultimate properties and performance of the manufactured article.

The thermo-mechanical history experienced by the material during processing is the result of interactions between resin properties, machine design, and operating conditions. In order to elucidate the details of the history, it is useful to employ mathematical models or computer simulations of the process. Moreover, it is important, if such models and simulations are to be useful, that appropriate information be used regarding material properties and boundary and initial conditions.

0097–6156/95/0603–0082$15.00/0

Thus, successful simulation of the process depends substantially on the availability of models describing material behavior under process conditions. Many of the important processing properties of polymers and models describing them have been the subject of earlier reviews (*1,2*). This chapter updates these reviews with emphasis on aspects relevant to simulation of injection molding of thermoplastics in complex-shaped cavities. Issues associated with crystallization kinetics and the coupling of crystallization kinetics with the solution of the transport equations are given special consideration. It should be possible to extend these aspects to the simulation of injection molding of thermosetting systems, since the kinetics and energetics of both crystallization and thermoset cure may be coupled similarly with the transport equations. Since rheological phenomena are treated extensively in many references, rheological aspects are not considered in this chapter.

One of the important performance criteria for plastics articles, especially those manufactured by injection molding, relates to the specifications regarding shrinkage and warpage limits. Both shrinkage and warpage are closely related to frozen thermal stresses and orientation in the molded article. Therefore, the last section of the chapter deals with a simple analysis of thermal stresses in molded articles and the effect of pressure on these thermal stresses. Finally, we report on the application of inverse modeling to thermal stresses, with the objective of estimating the optimal temperature or pressure variation during the injection molding cycle in order to achieve a prescribed distribution of thermal stresses.

A Transport Model of Injection Molding (Computer Simulation)

The injection molding process involves injecting the molten polymer into the cavity (filling stage); maintaining the pressure to introduce additional melt into the cavity in order to compensate for shrinkage (packing stage); and cooling the melt until it is solidified (cooling stage). The injection molding process is characterized by its transient nature and by the non-Newtonian, viscoelastic, non-isothermal flow accompanied by crystallization and solidification. Furthermore, the process involves free surface flow through channels and molds of irregular geometries. All these complexities affect the final properties of the molded article.

Computer simulation provides information regarding thermal and flow phenomena that prevail during the injection molding process. Models differ in the simplifying assumptions that are made and in the methods of solution. In most cases, simulations have employed 2-D solutions by using the Hele-Shaw approximation (*3*). This approximation is valid for thin mold cavities with large aspect ratios, and in which out-of-plane flows may be neglected. Hele-Shaw flow analysis was employed by Tadmor et al (*4*), Kamal et al (*5,6*), and Ryan and Chung (*7*) for simple geometries. In order to deal with the fountain flow near the melt front (*8*), a more elaborate filling model for simple geometries was proposed by Kamal et al (*9*).

The classical finite difference method (FDM) is useful for the analysis of injection molding in simple geometries. On the other hand, the finite element method (FEM) is usually employed in the simulation of molding in complex geometries. However, this could involve a sacrifice in the degree of physical detail and the complexity in coding. Hieber and Shen (*10*) employed a combined finite element/finite difference method for the solution of 2-D flow in arbitrary geometries. Couniot and Crochet (*11*) presented a finite-element based model for filling of a thin planar mold cavity. Following the advances in automatic grid generation (*12,13*) and in the method of coordinate mapping, namely Boundary Fitted Curvilinear Coordinates (BFCC), it has become possible to use the finite difference method for the treatment of complex geometries. Kamal and Papathanasiou (*14,15*) and Subbiah et al (*16*) proposed models based on the BFCC-FDM using a moving grid.

The use of Hele-Shaw flow for simplifying the 3-D equations to 2-D is valid only in the region far from the moving front. However, for many applications where the flow in the region of free surface is important, the Hele-Shaw approximation is inadequate due to ignoring the fountain flow in that region. Full 3-D models have been avoided in injection molding simulation due to the complexity of the resulting model and to the requirement of intensive computing resources. As a compromise, Friedrichs et al (*17*) proposed a hybrid 2-D/3-D technique. This approach solves the 3-D fluid flow equations only in the fountain flow region, while the 2-D Hele-Shaw formulation is employed behind the flow front region. However, the authors did not consider the non-isothermal and crystallization phenomena that prevail in the process.

In order to study the effect of fountain flow in the melt region and to provide accurate information regarding temperature and crystallinity distributions, we have developed a two-stage model. The model is based on the so-called 2½-D analysis that employs a 2-D flow analysis and a solution to the 3-D energy equation. In the first stage of the analysis, the fountain flow information is calculated from the solution of relevant equations in the thickness plane. Then, in the next stage, the analysis is carried out in planes parallel to the plane of the cavity. Therefore, it is possible to combine the two analyses to obtain a reasonable description of the overall flow field in the plane of the cavity, including fountain flow. The incorporation of the fountain flow is critical for the description of the temperature field, both near the flow front and near the mold wall. In particular, the evolution of crystallinity at the surface of the molding is strongly influenced by the fountain flow. This is achieved by coupling the fountain flow and the crystallization kinetics with the energy equation, as will be described later.

For the sake of simplicity, the present treatment deals with the filling and cooling stages only. Thus the role of pressure is ignored. Many of the results presented here will refer to the geometry of the mold shown in Figure 1. The thickness direction is denoted by z, the width direction by y, and the length direction is x, which is also the main flow direction.

Gapwise Direction. The gapwise analysis is conducted first in order to obtain the information regarding fountain flow. A reasonable approach is to carry out the analysis in the symmetry plane where the flow in the width direction is absent. The pressure in the analysis is decoupled from the solution of the momentum equations by using the vorticity (ω) - stream function (ψ) formulation:

$$\nabla^2 \psi = -\omega \tag{1}$$

$$\rho\left(\frac{\partial \omega}{\partial t} + u\frac{\partial \omega}{\partial x} + w\frac{\partial \omega}{\partial z}\right) = \nabla(\mu\nabla\omega) + Sc_\omega \tag{2}$$

u and w are the velocities in the x- and z-direction, respectively, and they are defined in terms of the stream function ψ as follows:

$$u = \frac{\partial \psi}{\partial z}$$
$$w = -\frac{\partial \psi}{\partial x} \tag{3}$$

and vorticity ω is defined as:

$$\omega = \frac{\partial w}{\partial x} - \frac{\partial u}{\partial z} \tag{4}$$

The source term Sc_ω in Equation 2 compensates for the contributions of elastic stresses, where the stress tensor (**T**) is decomposed into viscous (μ**D**) and elastic (**S**) parts. Decomposition of the stress tensor, as has been suggested by Perera and Walters (*18*), is a preferable stabilizing factor for the numerical solution that causes the equation to become diffusion-dominant. Sc_ω is, therefore, expressed as follows:

$$Sc_\omega = \frac{\partial(S^{zz} - S^{xx})}{\partial x \partial z} + \frac{\partial S^{xz}}{\partial x^2} - \frac{\partial S^{xz}}{\partial z^2} \tag{5}$$

where **S** = **T** - μ**D**. In this stage of the analysis, the contribution of stresses in the width direction is assumed to be negligible due to the large width to thickness ratio. The stress tensor **T** can be calculated from the constitutive models of viscoelastic fluids such as the White-Metzner equation. However, to simplify the analysis, a power law model is employed in this treatment, where the source Sc_ω vanishes. Thus, the viscosity dependency on shear rate is as follows:

$$\mu = Ae^{\frac{\Delta E}{RT}}\left(\frac{\Pi}{2}\right)^{\frac{n-1}{2}} \tag{6}$$

where n is the power law index and Π is the second invariant of the rate of deformation tensor.

Since the temperature gradient in the y-direction vanishes at the symmetry plane, the energy equation is simplified into the following 2-D form:

$$\rho . C_P\left(\frac{\partial T}{\partial t} + u\frac{\partial T}{\partial x} + w\frac{\partial T}{\partial z}\right) = \frac{\partial}{\partial x}\left(k\frac{\partial T}{\partial x}\right) + \frac{\partial}{\partial z}\left(k\frac{\partial T}{\partial z}\right) + \Phi + \rho.\Delta H_L\frac{D\chi}{Dt} \tag{7}$$

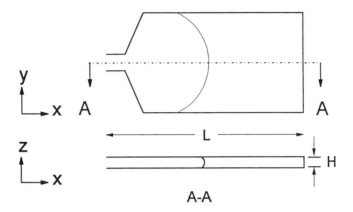

Figure 1.   Sketch of the mold cavity.

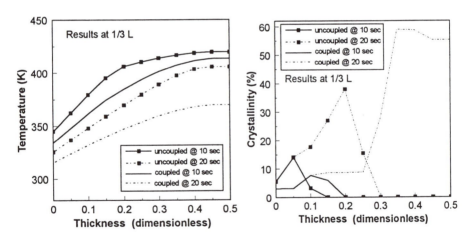

Figure 2.   Coupling effect on distribution of temperature and crystallinity across the thickness. Location 0 is cavity surface and 0.5 is the cavity center.

where Φ represents the dissipation function and ΔH_L is the latent heat of fusion. The rate of crystallization is expressed using the differential form of the non-isothermal Nakamura model (*19*) which will be discussed at length later.

$$\frac{D\chi}{Dt} = \chi_\infty \, m K(T)(1-a)\left[\ln\left(\frac{1}{1-a}\right)\right]^{\left(\frac{m-1}{m}\right)} \tag{8}$$

where

$$\frac{D\chi}{Dt} = \frac{\partial\chi}{\partial t} + u\frac{\partial\chi}{\partial x} + w\frac{\partial\chi}{\partial z}, \tag{9}$$

$K(T)$ is a rate constant; m is Avrami's constant; a is the relative crystallinity which is defined as χ/χ_∞; and χ_∞ is the ultimate crystallinity for the polymer.

Some models employ two equations to describe the energy balance: one for the liquid phase and the other for the solid phase (*20*). However, it is sufficient to describe the process using one equation (Equation 7) by properly coupling it with the mass balance. The coupling of crystallization kinetics and crystallinty mass balance (Equations 8 and 9) with the energy balance (Equation 7) together with the overall mass and momentum balances in the system (Equations 1 and 2) has a profound influence on the temperature and crystallinity fields. In general, the existing models of the injection molding process have not coupled crystallization kinetics with the energy equation. The model described here provides a distinct advantage regarding the analysis and prediction of crystallinity.

Figure 2 shows predictions of temperature and crystallinity distributions for the cases with and without coupling of crystallinity and the energy equation. It is obvious that there is a large difference, which suggests that the uncoupled solution does not provide satisfactory predictions of temperature and crystallinty profiles. The main deficiency of the uncoupled solution is due to the fact that the specific heat data are obtained at one cooling rate. However, crystallization during injection molding takes place over a broad range of cooling rates. The coupled solution allows for such variation in the cooling rates. Moreover, coupling of the heat of crystallization in the energy equation should provide a more realistic prediction of the temperature distribution during solidification.

In-Plane Direction. The analysis in the gapwise plane provides the information on the fountain flow needed for the second stage of analysis. The velocity w in the z-direction (fountain flow) is obtained and described by the function

$$w = g(d, z, X_{fs}) \tag{10}$$

where d is the normal distance from the free surface, and X_{fs} is the location of the free surface (i.e. time). The flow field in the in-plane is calculated using the 2-D model as in Equations 1 and 2, where the z-dimension is replaced by y and the velocity w is replaced by the velocity in the y-direction v. The average velocity plane is chosen for the calculation of the flow field in order to obtain realistic information about the advancement of melt front. The source term in the vorticity equation involves, now, additional terms due to the significant influence of shear stress terms in the thickness direction.

$$Sc_\omega = \frac{\partial^2\left(S^{yy} - S^{xx}\right)}{\partial x \partial y} + \frac{\partial^2 S^{xy}}{\partial x^2} - \frac{\partial^2 S^{xy}}{\partial y^2} + \left(\frac{\partial^2 \tau^{yz}}{\partial z \partial x} - \frac{\partial^2 \tau^{xz}}{\partial z \partial y}\right) \qquad (11)$$

The same power law form for viscosity is used as in Equation 5. The second invariant in Equation 5 and the z-direction shear stresses in Equation 11 contain the velocity gradients $\langle \partial u / \partial z \rangle$ and $\langle \partial v / \partial z \rangle$. These gradients are approximated using the assumption of fully developed power-law velocity profiles. However, in the fountain flow region near the melt front, the fountain flow velocity distribution given by Equation 10 is used in the convection term of the 3-D energy equation. Thus, the energy equation becomes:

$$\rho . C_P \left(\frac{\partial T}{\partial t} + u \frac{\partial T}{\partial x} + v \frac{\partial T}{\partial y}\right) = \frac{\partial}{\partial x}\left(k \frac{\partial T}{\partial x}\right) + \frac{\partial}{\partial y}\left(k \frac{\partial T}{\partial y}\right) + \frac{\partial}{\partial z}\left(k \frac{\partial T}{\partial z}\right) +$$

$$- \rho . C_P \left(w \frac{\partial T}{\partial z}\right) + \Phi + \rho \Delta H_L \frac{D\chi}{Dt} \qquad (12)$$

In the cooling stage, the velocities are set to zero, and Equations 12 and 8 are coupled and solved.

The importance of incorporation of the fountain flow effect is illustrated in the comparison shown in Figure 3. This figure shows that although the temperature distributions are indistinguishable at the end of cooling, the crystallinity distributions, which are influenced by the thermal history, are quite different if the fountain flow is not included. Note also that although the temperature distributions continue to change after 10 seconds, yet the crystallinity does not change since the temperature reached levels at which crystallization does take place.

Grid Transformation. The filling of irregular shaped cavities requires a numerical scheme that can cope with the complexity of both the geometry and the process. A moving boundary-fitting grid is used for the purpose, and a transformation of the grid to a uniform square grid is implemented. The finite difference method is, then, used for the discretization of the governing equations.

The grid is generated at each time step and the governing equations are transformed using the Boundary-Fitted Curvilinear Coordinate (BFCC) method. A typical general equation describing the variable ϕ which includes convection, diffusion and source terms, can be transformed from the Cartesian coordinate system (x,y) to curvilinear coordinate system (ξ,η) to yield the following equation.

$$\rho\left(\frac{\partial\phi}{\partial t}+\frac{(u-x_t)}{J}\left(y_\eta\phi_\xi-y_\xi\phi_\eta\right)+\frac{(v-y_t)}{J}\left(x_\xi\phi_\eta-x_\eta\phi_\xi\right)\right)=$$

$$\frac{\mu\left(\alpha\phi_{\xi\xi}-2\beta\phi_{\eta\xi}+\gamma\phi_{\eta\eta}\right)}{J^2}+\mu\nabla^2\xi\phi_\xi+\mu\nabla^2\eta\phi_\eta+S_c \tag{13}$$

where

$$\gamma=\left(x_\xi\right)^2+\left(y_\xi\right)^2$$

$$\alpha=\left(x_\eta\right)^2+\left(y_\eta\right)^2$$

$$\beta=x_\xi x_\eta+y_\xi y_\eta$$

and the Jacobean of transformation is defined as:

$$J=x_\xi y_\eta-x_\eta y_\xi$$

The indices represent a partial derivation with respect to the index; x_t, y_t represent grid velocities in the x- and y-directions.

Boundary Conditions. In the solution of the flow field in both planes, gapwise and in-plane, the stream function boundary conditions are similar. Constant streamline values are present at the walls of the mold. At the inlet, streamline values are calculated from the assumed fully developed profile of velocity. Since, the moving free surface is perpendicular to the stream lines, the normal gradient of the stream function at the free surface vanishes in order to satisfy the zero shear stress condition. In other words,

$$\psi\big|_{\text{mold walls}}=\text{Constant}$$

$$\frac{\partial\psi}{\partial n}\bigg|_{\text{free surface}}=0 \tag{14}$$

The no-slip condition for the vorticity equation is satisfied on the solid walls of the cavity. This is achieved by using the basic definition of the vorticity.

$$\omega\big|_{\text{wall}}=\left(\frac{\partial v}{\partial x}-\frac{\partial u}{\partial y}\right)_{\text{wall}} \tag{15}$$

The boundary condition in Equation 15 is expressed in terms of velocities so that any solidification that may occur is taken into consideration in the solution.

The thermal boundary condition corresponds to a constant coolant temperature. The flux is represented using the overall convective heat transfer coefficient h as follows:

$$\left(k\frac{\partial T}{\partial n}\right)_w = h\left(T_w - T_{coolant}\right) \tag{16}$$

The heat transfer coefficient will be discussed in more detail in the next section. The temperature at the melt front is set equal to the temperature of the front at mid-plane to further account for fountain flow. When the melt reaches the far end of the cavity, the boundary condition discussed in Equation 16 is employed on the free surface.

Heat Transfer Coefficient. For most applications, convective heat transfer coefficients are calculated from empirical equations obtained by correlating experimental data with the aid of dimensional analysis (*21*). Varying the Reynolds number generally allows for correlations between equipment of different sizes; varying the Prandtl number allows for the extrapolation to other liquids. The empirical correlations take the following forms:

Long Ducts: $\text{Nu} = \phi(\text{Re})\Phi(\text{Pr})$

Short Ducts: $\text{Nu} = \phi(\text{Re})\Phi(\text{Pr})f(x/D_H)$ (17)

where Re is Reynolds Number; Pr is Prandtl Number; x represents the position along duct; and D_H represents the hydraulic diameter. A simple empirical correlation presented by Sieder and Tate (*22*) which has been widely used to correlate experimental results for liquids in tubes can be written as follows: (all properties are based on the bulk temperature.)

$$Nu_{DH}\left(\frac{\mu_s}{\mu_b}\right)^{0.14} = 1.86\left(\text{Re}\,\text{Pr}\frac{D}{L}\right)^{\frac{1}{3}} \tag{18}$$

This equation can be applied when the surface temperature is uniform and in the range of $0.48 < \text{Pr} < 16,700$ and $0.0044 < (\mu_b/\mu_s) < 9.75$. It is recommended to use this equation only when $(\text{Re}_{DH}\,\text{Pr}\,D/L)^{0.33}(\mu_b/\mu_s)^{0.14}$ is larger then 2. It is important to note, when using the above equations for design purposes, that the error may be as much as ± 30 per cent for streamline conditions.

Charm and Merrill (*23*) derived the following equation for power law fluids:

$$\frac{hD_H}{k} = Nu_{DH} = 2.0(\text{Gz})^{1/3}\left[\frac{K_b}{K_W}\frac{(3n+1)}{2(3n-1)}\right]^{0.14} \tag{19}$$

where Gz is the Graetz number, n is the power law index and K refers to the power law constant.

More detailed treatment of heat transfer, including the estimation of Nusselt numbers and heat transfer coefficients in Newtonian and non-Newtonian flow systems, is available in texts by Skelland (*24*) and Wilkinson (*25*).

Insensitivity to melt temperature is not surprising when one considers the well-known heat transfer relationship for forced convection in laminar flow (*26*):

$$h\frac{D}{k} = 1.86\left(\operatorname{Re}\operatorname{Pr}\frac{D}{L}\right)^{0.33}\left(\frac{\mu_b}{\mu_s}\right)^{0.14}$$ (20)

This suggests the following proportionality:

$$h(T) \propto k\left(\frac{c_p}{k}\right)^{0.33}$$ (21)

Below 200°C, the right-hand side is practically constant for most polymers.

The numerical value of the heat transfer coefficient depends on the choice of a reference temperature. For flow over an object or plane surface, the temperature of the fluid far away from the solid is generally constant and is usually taken as the reference temperature. In a conduit, the temperature varies both in the direction of mass flow and in the direction of heat flow. At a given cross-section, the temperature in the center would be the most obvious choice for a reference temperature. However, in practice, this is difficult to measure and, therefore, the average fluid bulk temperature, T_b, is customarily used as the reference. In injection molding, T_b, the mixing cup temperature, is not practical for on-line control. Therefore, several other temperature measurements have been selected for comparison on the basis of predictive value.

Kamal and co-workers (*27*) have carried out temperature and heat flux measurements at strategically located positions in the injection molding system. On the basis of these measurements, they found that the injection velocity at the gate plays an important role in determination of the overall heat transfer coefficient in the mold. Polyethylene and polystyrene resins were used in the study.

The heat transfer coefficients calculated for polyethylene based on surface temperature T_s are described by the following correlation:

$$h\left(\frac{\text{Btu}}{\text{ft}^2 \text{ h } °\text{F}}\right) = 232 \times Velocity^{0.160} \qquad \text{based on } (T_s - T_{mold})$$ (22)

$$Velocity \rangle 50 \text{ cm / s}$$

It should be pointed out that the above correlation is applicable for the molding arrangement used in the above mentioned study (*27*). However, it may be applicable

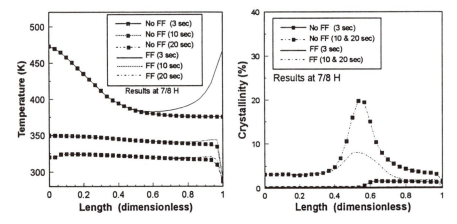

Figure 3.   Fountain flow effect on distribution of temperature and crystallinity along the cavity. Length 0 is the inlet and 1 is the end wall.

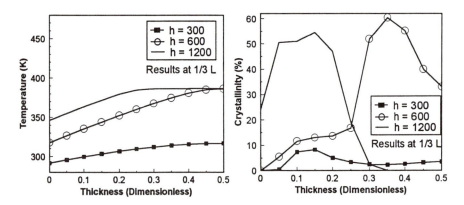

Figure 4.   Effect of heat transfer coefficient on crystallinity distribution across the thickness (20 sec.). Location 0 is cavity surface and 0.5 is the cavity center.

to other small cavities. For large complex moldings, it will be necessary to obtain separate correlations depending on the specific arrangement.

Figure 4 shows the strong impact of the value of the heat transfer coefficient on crystallinity distribution and indicates the importance of selecting the correct value for h.

Crystallization Kinetics

Avrami's Isothermal Model. Avrami (28) derived the following equation for the kinetics of phase change:

$$x(t) = 1 - \exp\left[-\frac{\rho_c}{\rho_e} \int_0^t v(t, \tau) \, \dot{N}(\tau) \, d\tau \right] \qquad (23)$$

where $x(t)$ is the actual fraction of the material transformed, $\dot{N}(\tau)$ is the nucleation frequency per unit volume, $v(t, \tau)$ is the corresponding volume of the growing center, and ρ_c and ρ_e are the densities of the crystalline and liquid phases, respectively. The simplifying assumptions usually made are that $\dot{N}(\tau)$ is constant and the growth process is linear. With these assumptions it is found that

$$v(t,\tau) = \sigma_m \, G^m(t - \tau)^m \qquad (24)$$

where σ_m is the shape factor: $(4\pi/3)$, $4\pi D$ or S for m = 3, 2, 1, respectively, where D is the thickness of a disk, and S is the cross-section of a rod. The rate of nucleation and the rate of growth relations were postulated by Hoffmann (29) and reviewed by Van Krevelen (30):

Rate of Nucleation: $\qquad N_o = \dot{N}_o \exp\left(-\frac{E_D}{RT}\right) \exp\left(-\frac{\Delta G_m}{kT}\right) \qquad (25)$

Rate of Growth: $\qquad G = G_o \exp\left(-\frac{E_D}{RT}\right) \exp\left(-\frac{\Delta G_m}{fT}\right) \qquad (26)$

The term $\exp(-E_D/RT)$ describes the temperature dependence of the rate of diffusive transport of the molecules in the melt. The ΔG_m term describes the free energy of a nucleus growing in an m-dimensional space, E_D is an activation energy for segmental jump rate in polymers.

These relations for the nucleation and growth processes may be introduced into the Avrami Equation 23 to yield:

$$x(t) = 1 - \exp\left(-kt^n\right) \qquad (27)$$

where n is the Avrami exponent, and k is the rate constant of crystallization, which can be approximated by (30):

$$k = \sigma_m \, \dot{N} \, G^m \left(\rho_c / \rho_e \right) \tag{28}$$

The time to reach 50 % conversion $t_{1/2}$, also termed the half-time, can easily be derived from Equation 27:

$$t_{1/2} = \left(\frac{\ln 2}{k} \right)^{\frac{1}{n}} \tag{29}$$

This relationship has been used to estimate the rate constant k. It should be noted that the Avrami Equation 27 was derived under the assumption that the phase transformation goes to completion. In polymers, complete crystallinity is rarely attained. Mandelkern (31) introduced the so-called generalized Avrami equation. It assumes the following form:

$$a\left(t, \, T \right) \equiv \frac{x(t)}{x_\infty(T)} = 1 - exp\left[-k(T) \, t^n \right] \tag{30}$$

where $x(t)$ is the actual fraction, $x_\infty(T)$ is the ultimate crystallinity and $a(t, \, T)$ is known as an effective fraction (or weight fraction) of the crystallized polymer at time t with respect to the maximum fraction achievable at temperature T. In Equation 30, $k(T)$ is the rate constant and n is the Avrami exponent.

Both the Avrami equation and the generalized Avrami equation are derived for isothermal conditions and assume that the characteristic kinetics of phase change remain the same during crystallization (isokinetic).

Nakamura's Non-Isothermal Model.   In polymer processing operations such as extrusion, injection molding and fiber spinning, the temperature varies with time and space and the polymer crystallizes under non-isothermal conditions. Therefore, studies of non-isothermal crystallization kinetics are quite important for optimizing processing conditions and obtaining required product properties.

Ziabicki (32) and Kamal and Chu (37) derived kinetic equations based on Avrami's theory and assumed the non-isothermal crystallization process may be treated as a sequence of isothermal crystallization steps, neglecting secondary crystallization. Nakamura et. al (19), based on the assumption that the growth of a crystal ceases when that crystal impinges on a neighboring one, obtained the following equation for non-isothermal crystallization:

$$a(t) = 1 - exp\left[-\left(\int_0^t K(T)dt \right)^n \right] \tag{31}$$

where n is the Avrami exponent determined from the isothermal crystallization data, and $K(T)$ is related to $k(T)$ as follows:

$$K(T) = \left[k(T)\right]^{\frac{1}{n}} \tag{32}$$

The assumption of the isokinetic condition employed by Nakamura is not likely to hold. Nakamura's equation of crystallization of polymers predicts fairly well the weight fraction $a(t)$ in non-isothermal crystallization only in the neighborhood of the half-time of the primary process. As in the Avrami equation, the Avrami exponent n is often found fractional and not constant in Nakamura's equation.

Tobin's Isothermal Model. The generalized Avrami equation of phase transition kinetics, Equation 30, can be also expressed in the following form:

$$\ln\left[\frac{1}{1 - a(t)}\right] = k(T)t^n \tag{33}$$

where n assumes values between 1 and 4 and is not necessarily integer. If $a(t)$ from Equation 33 is plotted against t, the resultant curve is sigmoidal, with $a(t=0)=0$ and $a(t=\infty)=1$. Experimental phase transition kinetics curves invariably have this form. A number of researchers have tried to use the Avrami equation to fit experimental data obtained from crystallizing polymers. A fit could be obtained only below 10-30% conversion. Above this, Equation 33 predicts fractional conversions which are too high at a given time. In other words, the generalized Avrami equation is not representative, in principle, of the actual crystallization process when the complete range of conversion is considered. So Tobin and co-workers (*33,34*) pointed out that the phase transition problem must be cast into the form of the Volterra integral equation:

$$\frac{a(t)}{1 - a(t)} = kN\,t^n + k\,I^*\int_0^t (t - W)^n \left[1 - a(W)\right]dW \tag{34}$$

where constant k contains nucleation and growth parameters, n is an integer whose value depends on the nucleation mechanism and the form of crystal growth, N is the initial number of heterogeneous nuclei and I^* is the rate of homogeneous nucleation. In fact, Tobin (*35*) derived Equation 34 for two separate cases: heterogeneous nucleation and growth

$$\frac{a(t)}{1 - a(t)} = kN\,t^n \tag{35}$$

and homogeneous nucleation and growth

$$\frac{a(t)}{1 - a(t)} = kI^* \int_0^t (t - W)^n [1 - a(W)] dW \tag{36}$$

For small $a(t)$,

$$\ln\left(\frac{1}{1 - a(t)}\right) \approx \frac{a(t)}{1 - a(t)} \tag{37}$$

and both equations, Avrami's Equation 33 and Tobin's Equation 35 agree with each other. The integral Equation 36 can be also reduced by zeroth-order solution to Avrami's form, Equation 33. This means that Tobin's model contains Avrami's model as a zeroth-order solution of Volterra's integral equation. As a first approximation, with respect to time, Tobin's equation gives the same result as Avrami's equation. At long times, however, Avrami's $a(t)$ approaches unity faster than in Tobin's $a(t)$. This is considered as an advantage for Tobin's model. However, as is properly concluded by Eder et. al (36), the slower approach of $a(t)$ to unity at large values of time, as predicted by Tobin, is not a consequence of the model. In Eder's opinion, this slower approach to unity points to the presence of an activation time spectrum. This can be attributed to "secondary" crystallization which may be just another way of expressing the same concept.

Figure 5 shows a comparison between Avrami's and Tobin's equations for the prediction of isothermal crystallization behavior of high density polyethylene. The experimental data are those represented by Kamal and Chu (37) for Sclair 2908 Polyethylene resin supplied by Dupont Canada. The two models yield comparable results. However, as indicated above, Avrami's model causes the melt to reach ultimate crystallinity faster than Tobin's model. It should be noted, however, that Tobin's model predicts higher rates of crystallization in the early stages of the process.

Tobin's Non-Isothermal Model. Choe and Lee (38) combined both elements of Tobin's model: (a) heterogeneous nucleation and growth and (b) homogeneous nucleation and growth. Since the development of a crystalline phase in the polymer melt includes the above two competing nucleation mechanisms, they assumed that the crystallization behavior may be described as a linear combination of Equations 35 and 36. It is to be noted that Equation 34 corresponds to two different nucleation mechanisms occurring in parallel. This concept is physically justifiable and was successfully applied earlier by Malkin and co-worker (39) to the macrokinetic crystallization in polymers. Equation 34 is not convenient for use to study non-isothermal crystallization. The differential form of this equation is more adequate to analyze the non-isothermal process. Differentiating Equation 34 with respect to time t yields:

$$\frac{da(t)}{dt} = \frac{da_1(t)}{dt} + \frac{da_2(t)}{dt} \tag{38}$$

where

$$\frac{da_1(t)}{dt} = knN\, t^{n-1} \left[1 - a(t) \right]^2 \tag{39}$$

and

$$\frac{da_2(t)}{dt} = kn\, I^* \left[1 - a(t) \right]^2 \int_0^t (t - W)^{n-1} \left[1 - a(W) \right] dW \tag{40}$$

It should be pointed out that $da_1(t)/dt$ gives the rate of variation of relative crystallinity due to the heterogeneous nucleation and growth processes, while the $da_2(t)/dt$ is the rate of variation of weight fraction crystallinity due to the homogeneous nucleation and growth processes. Using the Hoffmann and Van Kravelen growth rates of spherulites, and substituting them into Equations 39 and 40, Choe and Lee obtained the following expression for the non-isothermal crystallization:

$$\frac{da(t)}{dt} = \frac{da_1(t)}{dt} + \frac{da_1(t)}{dt}$$

$$= k_1 \exp\left(-\frac{3E_D}{RT} \right) \exp\left(-\frac{3\Psi_1 \overset{\circ}{T}_m}{T\Delta T} \right) t^2 \left[1 - a(t) \right]^2 +$$

$$k_2 \exp\left(-\frac{4E_D}{RT} \right) \exp\left(-\frac{(3\Psi_1 + \Psi_2)\overset{\circ}{T}_m}{T\Delta T} \right) \times$$

$$\left[1 - a(t) \right]^2 \int_0^t (t - W)^2 \left[1 - a(W) \right] dW \tag{41}$$

where $\overset{\circ}{T}_m$ is an equilibrium melting temperature, E_D is the activation energy of diffusion of the crystallizing segments across the phase boundary, $k_1 = 4\pi I_0 V_0^3$, Ψ_1 is a constant related to the free energy of formation of a critical nucleus, Ψ_2 is a constant related to the free energy of formation of a growth embryo, and ΔT is supercooling. According to Choe and Lee, in the region near the melting point, the heterogeneous nucleation and growth process controls the rate of crystallization. As Malkin first pointed out (40), this is why the crystallization process is self-accelerating. At high crystallization temperatures or at low cooling rates, it can be therefore assumed that

$$a_1(t) \gg a_2(t) \tag{42}$$

Under this assumption Equation 41 reduces to Equation 39. Usually the parameter Ψ_1 is expressed in terms of E_d and T_{max} from Ziabicki's relationship (41):

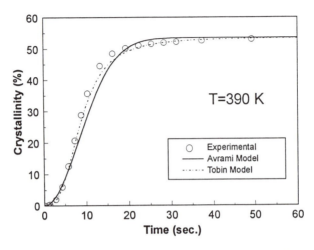

Figure 5.   Comparison between isothermal crystallization models of Avrami and Tobin and experimental data.

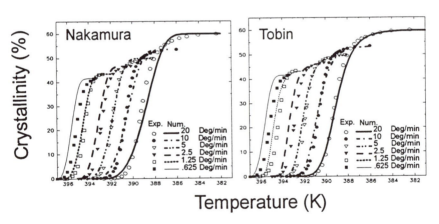

Figure 6.   Comparison between non-isothermal crystallization models of Avrami and Tobin and experimental data.

$$\Psi_l = E_d \frac{(T - T_{max})^2}{D^2} \tag{43}$$

where T_{max} is the temperature at which the growth rate of the spherulite is maximum. Once E_d is estimated, Ψ_1 can be determined from Equation 43.

When the semi-crystalline polymer is subjected to low temperature or high cooling rates, then the crystallization kinetics must be treated according to Equation 42, which includes both heterogeneous and homogeneous processes. It is also to be noted that Kim and co-workers (*42*) successfully applied the non-isothermal crystallization kinetics equation of the differential type (*38*) to analyze the crystallization characteristics of isotactic PP with and without a nucleating agent.

It is interesting to note that, under certain assumptions, the heterogeneous crystallization rate equation of Tobin becomes identical with an equation proposed by Kamal and Sourour (*43*) for the kinetics of cure of thermosetting polymers. This emphasizes again the similarity between the treatment of crystallization and cure in injection molding simulations.

Comparison between the non-isothermal crystallization predictions of the Tobin and Avrami models are shown in Figure 6 for the same polyethylene resin considered in Figure 5. Again, the Avrami (Nakamura) model predicts faster approach to the ultimate crystallinity especially at fast rates of cooling. Yet, at these rates, the initial rates of crystallization are higher for Tobin's model. Similar observation may be made regarding the predictions of crystalliniy distributions in injection molding by the Avrami and Tobin models, as shown in Figure 7. These results are obtained during the cooling stage at 10 and 20 s. after the beginning of filling.

Malkin's Macrokinetic Crystallization Equation.   Malkin and co-workers (*39*) suggested that the process of crystallization, in polymer processing, is self-accelerating. The larger is the crystal surface area, the higher will be the speed of crystallization. Later in the process, the space-filling problem becomes dominant. On the other hand, heterogeneity and disorder take place during the crystallization process. Intensive heat release is known to take place which may raise the temperature, causing the process rate to change. As a consequence of the heat release into the surrounding medium, the temperature distribution throughout the domain becomes non-uniform, which leads to substantially non-uniform levels of crystallinity.

Malkin et al. (*40*), proposed that the crystallization rate can be written as a linear combination of the two nucleation and growth processes as follows:

$$\dot{X}(T, X) = \dot{X}_1(T, X) + \dot{X}_2(T, X) \tag{44}$$

where X is the degree of crystallinity varying from zero to its equilibrium (ultimate or X_∞) value, $\dot{X}_1$ is the rate at which the degree of crystallinity varies as a result of the emergence of the primary nuclei and $\dot{X}_2$ is the rate of variation in the degree of crystallinity due to crystal growth. It can be shown that the non-isothermal Tobin model (Equations 38, 39 and 40), can be obtained from Malkin's model, Equation 44, as a special case. The Malkin model is formulated as follows:

$$X(t) = K(X_\infty - X)(1 + cX)$$ (45)

where K and c are generally temperature dependent. Equation 45 may be written in terms of relative crystallinity $a(t)$

$$a(t,T) \equiv \frac{X(T)}{X_\infty(T)}$$ (46)

as follows,

$$\dot{a}(t) = [A_1 + A_1 a](1 - a)$$ (47)

Constant A_1 corresponds to the initial rate at $a=0$, and A_2 corresponds to the auto-acceleration effect. Equation 47 shows that the rate of crystallization does not need to be equal to zero if the degree of crystallinity is identically zero in the molten polymer. This implies that primary nucleation plays an important role in the development of the overall crystallization process. It also reflects the fact that the nuclei of the new phase may be present or form instantaneously in a spontaneous, manner (for instance, under locally acting shearing forces) in the melt, which is not taken into account in models based on the Avrami equation.

The temperature dependence of constants A_1 and A_2 can be established from the measurements of crystallization rates. Commonly accepted forms satisfy the following relations:

$$A_1 = A_o \exp\left(-\frac{E_d}{RT}\right) \exp\left(\frac{-\psi_1 T_m^\circ}{T \Delta T}\right)$$ (48)

$$A_2 = K_o \exp\left(-\frac{(\psi_2 - \psi_1) T_m^\circ}{T \Delta T}\right)$$ (49)

where E_d is, as before, the activation energy of the segment transfer across the nucleus - melt boundary, R is the gas constant, T_m° is the equilibrium melting point, T is the current temperature and A_o, K_o, Ψ_1, Ψ_2 are constants. ΔT in Equations 48 and 49 is supercooling, i.e. $\Delta T = (T_m^\circ - T)$.

Equations 47 with 48 and 49 were verified using experimental data for PE, PP, PET, PP-oxide and polyurethane. The agreement of the theory with experiment was very good over a broad range of crystallinity values. In Equations 48 and 49 there are five parameters to be determined: E_d, Ψ_1, Ψ_2, A_o and K_o. Assuming that the activation energy E_d is known for the polymers and assuming that $\Psi_1 = \Psi_2 = \Psi$, the number of unknown parameters is reduced to three only. The activation energy E_d in polymers is usually taken to be 1500 cal/mole, as reported by Hoffmann et al (*44*) and Suzuki and Kovacs (*45*). Malkin and co-workers (*41*) applied the macrokinetic description of crystallization to the one-dimensional case for Nylon-6 melt by coupling the heat conduction equation with the crystallization kinetic equation, under the assumption that $\Psi_1 = \Psi_2 = \Delta$. Unknown parameters were determined by solving the inverse problem.

The Dependence of Rate Constant on Temperature. Crystallization rates in polymers are strongly dependent on temperature and stress (molecular orientation). The crystallization rate constant, K, is equal to zero above the melting temperature T_m and below the glass transition temperature T_g. In the intermediate region, it exhibits a maximum. The position and height of this maximum depend on the temperature of crystallization and degree of molecular orientation. The degree of crystallinity a is controlled by the maximum crystallization rate K_{max} and the increment of time Δt spent by the polymer in the vicinity of this maximum, that is, a is inversely proportional to the cooling rate dT/dt:

$$a \sim X_{max} \Delta t = \frac{K_{max} \Delta T}{(dT/dt)_{T_{max}}} \tag{50}$$

Ziabicki (*46*) has shown that, for uniaxial deformation, the effect of orientation can be represented by the following equation:

$$\left[t_{1/2}(T, f_a) \right]^{-1} = \left[t_{1/2}(T, 0) \right]^{-1} \exp\left[-4\ln(2)\frac{(T - T_{max})^2}{D^2} + A(T)f_a^2 + B(T)f_a^3 + ... \right] \tag{51}$$

where $t_{1/2}(T, f_a)$ and $t_{1/2}(T, 0)$ are the half-times for crystallization under molecular orientation and under no molecular orientation, respectively. The parameters A, B, etc. are characteristic of the polymer and are functions of temperature; f_a is an orientation function for the amorphous polymer before crystallization. In the special case where there is no orientation, i.e. $f_a = 0$, Equation 51 is reduced to an empirical formula for the temperature dependence of the crystallization half-times proposed earlier by Ziabicki (*32*):

$$\left(\frac{1}{t_{1/2}} \right) = \frac{1}{(t_{1/2})_{max}} \exp\left[-4\ln(2)\frac{(T - T_{max})^2}{D^2} \right] \tag{52}$$

where $(1/t_{1/2})_{max}$, T_{max} and D can be determined from experimental data. It should be pointed out that the predictions of crystallinity in injection molding are influenced substantially by the value selected for T_{max} in Equation 52.

Takayanagi and Kusomoto (*47*) have proposed the following equation for the dependency of the rate constant $K(t)$ on temperature:

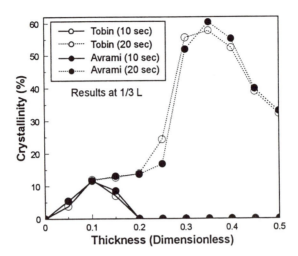

Figure 7.   Effect of crystallization model on crystallinity distribution across the thickness of the cavity. Location 0 is cavity surface and 0.5 is the cavity center.

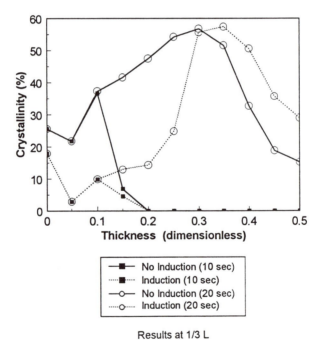

Figure 8.   Effect of induction time on temperature and crystallinity distribution across the thickness of the cavity. Location 0 is cavity surface and 0.5 is the cavity center.

$$\ln[K(t)] = A - \frac{BT}{(T - T_g + 51.6)^2} - \frac{CT_m}{T(T_m - T)}$$ (53)

where A, B, and C are constants and T_m is the melting point. Difficulties are encountered in fitting Equation 53 to experimental data.

Other Factors Influencing Crystallization Kinetics

Induction Time It is well known that crystallization does not always start spontaneously, but usually an induction period is observed before crystallization occurs. Sifleet (*48*) proposed the following equation to evaluate the effect of induction time in non-isothermal crystallization experiments:

$$\theta = \int_{T_0}^{T} \left(\frac{1}{t_I}\right) dt$$ (54)

where t_I is the isothermal induction time at temperature T, and T_0 is the initial temperature. Note that t_I is a function of T or the degree of supercooling. Crystallization in non-isothermal experiments starts when $\theta = 1$. Figure 8 shows a comparison between model predictions with and without inclusion of the induction time. It is seen that, except in the center of the cavity, ignoring induction time tends to yield an overprediction of crystallinity, as expected. The behavior of crystallinity at long times in the center of the cavity is probably the result of the balance of higher temperatures and lower cooling rates experienced when induction time is not included in the calculation. The higher temperatures are caused by the early release of heat of crystallization in that case.

Flow Effects. The effect of flow on crystallization kinetics is rather complex and there is need for accurate and practical models to explain the effects of shear rate, shear stress, and orientation on the kinetics of crystallization. Therefore, this subject will not be covered in detail in this work.

An empirical approach has been used by Hsiung and Cakmak (*49*) to estimate the effects of stress on induction time, t_I, and rate constant, K. They employed the following equations:

$$\log(S) = \log(S_b) + A(T - T_b)^2$$ (55)

$$T_b = T_{bq} + B\sigma$$ (56)

$$\log(S_b) = \log(S_{bq}) - C\sigma$$ (57)

In the above, S refers to either induction time t_I or the rate constant $K(t)$; A, B, and C are empirical constants; σ is the stress, and subscript q refers to the value of the variable (T_b or S_b) under quiescent conditions.

Eder and Janeschitz-Kriegl (50) proposed the following differential equation for growth and decay of a probability function P for the presence of a nucleation precursor:

$$\tau \dot{P} = \left(\frac{\dot{\gamma}}{\dot{\gamma}_a}\right)^2 (1-P) - P \qquad (58)$$

where $\dot{\gamma}$ is the shear rate under consideration, $\dot{\gamma}_a$ is the critical shear rate of activation, and τ is the relaxation time.

Pressure Effects. Pressure affects both the crystallization temperature T_m and the crystallization kinetics. The effect of pressure P on T_m is governed by the Clausius-Clapeyron equation:

$$\frac{dT_m}{dP} = T_m^0 \frac{\Delta V_m}{\Delta H_m} = \beta \qquad (59)$$

where T_m^0 is the crystallization temperature at zero pressure; ΔV_m and ΔH_m are the volume and enthalpy changes upon crystallization; and β is a constant. Van Krevelen (30) showed that for large pressure variations:

$$P - P_0 = a\left[\left(\frac{T_m}{T_m^0}\right)^c - 1\right] \qquad (60)$$

where a and c are constants. For polyethylene, $T_m^0 = 409$ K, $a \approx 3$ Kbar and $c = 4.5$. This suggests a rise in T_m of 20 K at 1 Kbar. Comparable changes in T_m are reported for polypropylene and polyamide 66 by Jing He (51).

According to data reported by He (51) for polypropylene, the effect of pressure on the Avrami exponent n is small. However, the pressure has a substantial impact on the crystallization rate constant K. According to these data, a doubling of pressure from 100 MPa to 200 MPa causes an increase in K equivalent to five orders of magnitude. Such a large effect would be very significant in injection molding, especially during the packing stage.

Residual Stresses in Injection Molding

During the flow of the melt in the filling and packing stages of injection molding of plastics and during non-uniform cooling of the material in the cooling stage until solidification, shear and normal stresses are generated in the sample. Part of these stresses relax during the process, but as a result of an increase in relaxation time at lower temperatures and change in elastic properties by solidification, part of these stresses are frozen and remain in the product as residual stresses. Residual stresses in molded articles may affect their quality by causing dimensional and shape instability, warpage, and highly anisotropic mechanical and optical properties (52,53). Residual stresses may be classified into two main groups: flow stresses and thermal stresses.

Flow stresses are generated during the filling and packing stages and include shear stresses and normal stress differences in shear flow and extensional stresses in extensional flow. Shear stresses are predominant in injection molding and they occur when adjacent layers of the polymer melt move at different velocities. Normal stress differences appear as a result of viscoelasticity of the melts and the large relaxation time of the polymeric chains. Extensional stresses are produced in fountain flow and wherever there is extensional flow in the regions of contraction and expansion in the mold. Attempts have been made to predict residual flow stresses together with velocity and pressure fields by solving the conservation equations simultaneously with appropriate constitutive equations (54).

Thermal stresses are generated as a result of inhomogeneous cooling of the material in the mold. These stresses remain in the product because of large changes of the elastic properties of the polymer with temperature during solidification.

If a body is cooled under no applied force, the process is called free quenching. When the temperature of a body is decreased by ΔT, the thermal contraction strain associated with this temperature drop is given by:

$$\varepsilon_T = \alpha \, \Delta T \tag{61}$$

where α is thermal expansion coefficient. If the body is allowed to deform freely, it will undergo a deformation equal to the thermal strain and no stress is generated. If part or all of the deformation is prevented, normal thermal stresses are imposed on the body. In the case of rapid cooling, the temperature of the body differs from one point to the other, therefore, during the same time, the temperature drop and thus the thermal strain is different for different elements of the body. To maintain the continuity of the body, the deformation of each element is restricted by the neighboring elements and consequently, thermal stresses are generated in the sample. For these stresses to persist in the absence of cooling, the elastic properties of the material that define its resistance to deformation, must change significantly with temperature during the cooling process.

Free Quenching. In case of free quenching of a one-dimensional sample with a temperature gradient in the y direction, the generated thermal stresses σ_{xx} and σ_{zz} are equal and are denoted by σ. Using the equations of thermoelasticity (55), the thermal stress in the layer y at time t may be related to the difference between the actual shrinkage $\delta(t)$ and the thermal strain as follows:

$$\sigma(y,t) = \left(\frac{E}{1-\nu}\right)\left[\delta(t) - \varepsilon_T(y,t)\right] \tag{62}$$

where E is the elastic modulus and v is the Poisson ratio of the material. Assuming that the elastic modulus in the melt is negligible compared to that of the solid state and using the equilibrium of forces in the solid part at each time, the shrinkage is obtained as follows:

$$\delta(t) = \frac{\int_{y_s}^{b} \varepsilon_T(y,t)dy}{(1-y_s)} \tag{63}$$

where b is half the thickness of the sample and y_s is the position of the solid-melt interface at time t.

Cooling Under Applied Pressure. In injection and compression molding of plastics, the polymer melt is cooled in a mold under applied pressure which varies with time $P(t)$. Using the equations of thermoelasticity, two models were developed for this case. In the first one developed by Titomanlio et al.(*56*), the sample is allowed to shrink in the mold. The thermal stress in the layer y at the time is given by:

$$\sigma(y,t) = \left(\frac{E}{1-v}\right)\left[\delta(t) + \eta(y) - \varepsilon_T(y,t) - \frac{v}{E}P(t)\right] \tag{64}$$

where $\eta(y)$ is the pressure strain tending to expand each solidifying layer.

In the second model developed by Brucato et al. (*57*), the problem was simplified by assuming that the sample keeps the dimensions of the mold before it is opened. For this case, the final stress distribution in the product may be calculated from the following expression:

$$\sigma(y) = \left(\frac{2v-1}{1-v}\right)\left[P(y) - \int_{0}^{b} P(y).dy\right] \tag{65}$$

where $P(y)$ may be obtained from the pressure history $P(t)$ and the transient position of the solid-melt interface $y_s(t)$. Note that in this model, the sample will be stress-free in the absence of pressure, whereas the first model gives the expression for free quenching in such a case.

Figure 9 shows the residual stress distribution calculated from the two models using the injection molding conditions of the experiments of Menges and Dierkes (*58*). The experimental measurements are shown in this figure by symbols. The free quenching stress profile consists of a compressive part close to the surface and a tensile part in the center (*53*). In the case of injection molding, the models show a tensile stress at the surface and in the center and a compressive region between them. The simplified model seems to agree well with the experiments aside from the surface where there are no experimental data. Tensile stresses on the surface have been reported experimentally (*59*).

Inverse Modeling. In a directly-posed or well-posed problem, the domain and boundaries of the problem, the governing and constitutive equations, the boundary and initial conditions, and the material properties are known. A problem in which any of this information is not available is an ill-posed or inverse problem. Experimental data are commonly used to obtain the unknown information. The difficulties associated with inverse problems are non-existence, non-uniqueness, and instability of the solution. In the processing of polymeric materials, several inverse problems of practical importance arise. Cohen (*60*) used the inverse formulation together with the method of characteristics to obtain the extruder die profile providing a uniform flow with a prescribed total flux.

Using previous direct models (*56,57*), Farhoudi and Kamal (*61,62*) developed an inverse model to calculate the pressure history or initial temperature of the melt in injection molding from measurements of residual stresses in the product. Equation 65 can be written as:

$$\underline{\sigma}(y) = \left(\frac{2v-1}{1-v}\right)\underline{P}[I - DY]$$

(66)

The components of vector $\underline{P}$ represent the pressure at different time steps during packing and cooling. The components of $\underline{\sigma}$ give the residual stress at various positions Y in the molded sample. The NxN tensor I is the unit tensor and DY is the symmetric NxN tensor containing DY's; N is the number of data points at which residual stress is known. Upon rearrangement, Equation 66 become

$$\underline{P}[I - DY] = \left(\frac{1-v}{2v-1}\right)\underline{\sigma}$$

(67)

Equation 67 may be inverted to solve for $\underline{P}$ or the sum of residuals of this equation may be minimized to obtain the optimum $\underline{P}$. The residual vector $\underline{R}$ may be defined as follows:

$$\underline{R} = \underline{P}[I - DY] - \left(\frac{1-v}{2v-1}\right)\underline{\sigma}$$

(68)

Farhoudi and Kamal tested their inverse method by the direct solution and experimental results of Menges and Dierkes (*58*). Figure 10 shows the results of the inverse method in case of unknown initial temperature distribution. The stress data were obtained from the direct simplified model (*57*) using a parabolic temperature profile. The actual profile was reproduced well with the help of a regularization method to insure uniqueness of the inverse solution. The pressure starts at zero at the end of filling (≈ 1.0 s.).

Figure 11 shows the pressure history calculated from the inverse model in case of unknown pressure history. The stress data were again calculated from the simplified model assuming a parabolic initial temperature profile in the melt. The pressure history obtained from the actual experimental data of Menges and Dierkes (*58*) is shown in

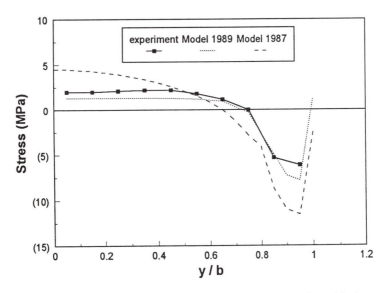

Figure 9.   Comparison between calculated and measured residual stress distributions.

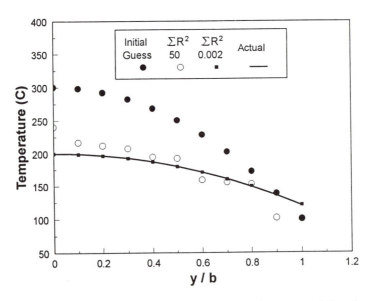

Figure 10.   Initial temperature profile predictions by inverse modeling (second order regularization parameter is equal to 0.08).

Figure 12. The dashed line shows a profile consistent with the reported holding pressure of 20 Mpa and holding time of 5 second and similar to the profile indicated by Titomanlio (*57*).

Conclusions

In injection molding, as in practically all plastics processing operations, the final product is obtained by solidification of the melt over a range of temperatures and cooling rates. As a result, the products obtained at the end of the process exhibit a distribution of crystallinity which varies widely from one point in the product to the other. Moreover, the variability of temperature, cooling rates, and material properties produce a distribution of thermal stresses in the product. The final distributions of crystallinity and thermal stresses in plastics products have a profound influence on the ultimate physical properties (e.g. mechanical, electrical, thermal properties, permeability etc.) and on the thermal and dimensional stability of the material (e.g. shrinkage and warpage). Therefore, a good understanding of the development of crystallinity and thermal stresses is of great practical significance.

This manuscript indicates the complexity of modeling plastics processing operations. This is particularly true for the injection molding process, which involves complex three-dimensional cavities and free surface, non-isothermal flow of rheologically complex melts. Furthermore, analysis of the solidification of partially crystalline polymers requires detailed information   regarding the temperature distributions throughout the injection molding cycle. In the above treatment, it has been shown that a complete analysis of the process must take into consideration the fountain flow near the free surface. Moreover, uncoupled treatment of crystallization leads to unsatisfactory results due to neglecting the effect of heat of crystallization on the temperature distribution and the fact that the uncoupled solution employs specific heat data obtained at only one cooling rate, while the solidification process occurs over a broad range of cooling rates. The results shown above also indicate the importance of kinetic models and the details regarding some of the crystallization kinetics parameters in predicting the crystallinity distribution in the final product.

The role of pressure in the development of thermal stresses cannot be ignored. Unfortunately, only limited data are available in the literature to evaluate models that consider the pressure effect. For example, there is need to evaluate whether the stress at the surface is always compressive, since some models predict a tensile stress under some conditions. The above treatment has shown that inverse modeling could provide a valuable tool for the optimization of thermal stresses, and possibly other characteristics such as the distributions of crystallinity in injection molded articles.

Further work is needed to elucidate the effects of flow stresses and pressure on crystallization kinetics in order to obtain a more complete and accurate analysis of crystallinity development during injection molding. With regard to the analysis of thermal stresses, there is need to extend the above one-dimensional analysis to the case of more realistic three-dimensional temperature distributions.

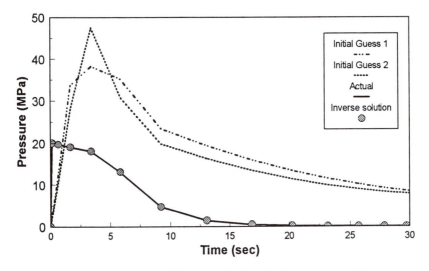

Figure 11.   Pressure history predictions by the inverse model.

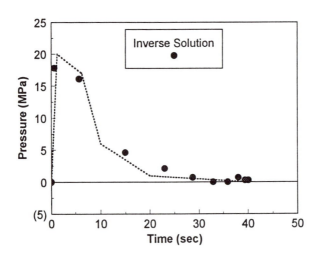

Figure 12.   Comparison between experimental data and inverse model predictions.

Acknowledgement

The authors would like to thank Ms. Yalda Farhoudi and Mr. Mazen Samara for their help in the preparation of the manuscript. Financial support was received in the course of this work from the Natural Science and Engineering Research Council of Canada and the Ministry of Education, Government of Quebec.

Literature Cited

1.   Kamal, M. R.; Ryan, M. E., *Models of Material Behavior: Fundamental of Computer Modeling for Polymer Processing* ; C. L. Tucker III, ed.; Hanser Publishers: N.Y., 1989; pp. 7-68.

2.   Kamal, M. R.; Ryan, M. E., *Adv. Polym. Technol.* **1984**, 4, 323-334.

3.   Richardson, S., *J. of Fluid Mech.* **1972**, 56 (4), 609-618.

4.   Tadmor, Z.; Broyer, E.; Gulfinger C., *Polym. Eng. Sci.* **1974**, 14, 660.

5.   Kamal, M. R.; Kuo, Y.; Doan P. H., *Polym. Eng. Sci.* **1975**, 15, 863.

6.   Kao, Y. ; Kamal, M. R., *AIChE J.* **1976**, 22, 661.

7.   Ryan, M. E.; Chung, T. S., *Polym. Eng. Sci.* **1980**, 20, 642.

8.   Schmidt, R. L., *Polym. Eng. Sci.* **1974**, 14, 797.

9.   Kamal, M. R.; Chu, Ed.; Lafleur P. G.; Ryan M. E., *Polym. Eng. Sci.* **1986**, 26, 190.

10. Hieber, C. A.; Shen, S. F., *J. non-Newtonian Fluid Mech.* **1980**, 7 1-32.

11. Counoit, A.; Crochet, M. J., *The Annual Technical Workshop of the Center for Composite Materials*, University of Delaware, Newark, Delaware, January 1987.

12. Thompson, J. F.; Thames, F. C.; Mastin, C. W., *J. Comp. Phys.* **1974**, 15, 299.

13. Thompson, J. F.; Warsi, Z. U. A.; Mastin, C. W., *Numerical Grid Generation: Foundations and Applications,* North Holland Publishers, 1985.

14. Papathansiou, T. D.; Kamal, M. R., *Polym. Eng. Sci.* **1993**, 33 (7), 400-409.

15. Papathansiou, T. D.; Kamal, M. R., *Polym. Eng. Sci.* **1993**, 33 (7), 410-417.

16. Subbiah, S.; Trafford, D. L.; Guceri, S. I., *Int. J. Heat Mass Transfer* **1989**, 32 (3), 415-434 .

17. Friedrichs, B.; Guceri, S. I., *J. Non-Newtonian Fluid Mech.* **1993**, 49, 141-173.

18. Perera, M. G. N.; Walters, K., *J. Non-Newtonian Fluid Mech.* **1977**, 2, 191.

19. Nakamura, K.; Watanabe, T.; Katayama, K., *J. Appl. Polym. Sci.* **1972**, 16, 1077-1091.

20. Chen, B. S.; Liu, W. H., *Polym. Eng. Sci.*   **1989**, 29 (15), 1039.

21. Bohn; Mark, S.; Frank, Kreith, *Principles of Heat Transfer*, 4th ed., Harper and Row Publishers, New York, 1986.

22. Sieder, E. N.; Tate G. E., *Ind. Eng. Chem.* **1936**, 28, 1429.

23. Charm, S. E.; Merrill, E. W., *Food Research* **1959**, 24, 319.

24. Skelland, A. H. P., *Non-Newtonian Flow and Heat Transfer*, John Wiley & Sons, Inc., N.Y., 1967.

25. Wilkinson, W. L., *Non-Newtonian Fluids: Fluid Mechanics Mixing and Heat Transfer*, Pergamon Press, London, 1960.

26. Hulatt, M.; Wilkinson, W. L, *Polym. Eng. Sci.*   **1978**, 18, 1148-1154.

27.   Kamal, M. R.; Mutel, A. T.; Garcia-Rejon A.; Salloum, G, *ANTEC proceedings* **1991**, 37, 483-487.

28.   Avrami, M., *J. Chem. Phys.* **1939**, Vol. 7, 1103; **1940**, Vol. 8, 212; **1941**, Vol. 9, 177.

29.   Hoffmann, J. D.; *J. Chem. Phys.* **1958**, 28, 1192.

30.   Van Krevelen, D. W., *Chimia* **1978**, 32, 279.

31.   Mandelkern, L., *J. Appl. Phys.* **1955**, 26, 443.

32.   Ziabicki, A., *Colloid Polym. Sci.* **1974**, 252, 207.

33.   Tobin, M. C., *J. Polym. Sci. Phys. ed.* **1974**, 12, 399.

34.   Tobin, M. C.; Fowler, J. P.; Hoffmann, H. A.; Sauer, C. W., *J. Amer. Chem. Soc.* **1954**, 76, 3249.

35.   Tobin, M. C., *J. Polym. Sci. Phys. ed.* **1976**, 14, 2253.

36.   Eder, G.; Janeschitz-Kriegl, H.; Liedouer, S., *Prog. Polym. Sci.* **1990**, 15, 629.

37.   Kamal, M. R.; Chu, E., *Polym. Eng. Sci.* **1983**, 23, 27.

38.   Choe, C. R.; Lee, K. H., *Polym. Eng. Sci.* **1989**, 29, 801.

39.   Malkin, A. Ya.; Beghishev, V. P.; Keapin, I. A.; Bolgov, S. A., *Polym. Eng. Sci.* **1984**, 24, 1396.

40.   Malkin, A. Ya.; Beghishev, V. P.; Keapin, I. A.; Adrianova, Z. S., *Polym. Eng. Sci.* **1984**, 24, 1402.

41.   Ziabicki, A., *Fundamentals of Fiber Formation*, J. Wiley & Sons, 1976.

42.   Kim, Y. C.; Kim, C. Y.; Kim, S. C., *Polym. Eng. Sci.* **1991**, 31, 1009.

43.   Kamal, M. R.; Sourour, S., *Polym. Eng. Sci.* **1973**, 13, 59-64.

44.   Hoffmann, J. D.; Davis, G. T.; Lounitzen, J. I., *Treatise on Solid State Chemistry: Crystalline and Non-Crystalline Solids*, vol. 3, Ch. 7, B. Hannay, ed., Plenum, N.Y., 1976.

45.   Suzuki, T.; Kovacs, A. J.; *Polymer J.* **1970**, 1, 82.

46.   Ziabicki, A., *J. Chem. Phys.* **1986**, 85, 1.

47.   Takayanagi, M.; Kusumoto, T., *Kagyo kagaku Zasshi* **1959**, 62, 587.

48.   Sifleet, W., *Polym. Eng. Sci.* **1973**, 13 (1), 10.

49.   Hsiung, C. R.; Cakmak, M., *Polym. Eng. Sci.* **1991**, 31, 1372.

50.   Eder, G.; Janeschitz-Kriegl, H., *Colloid Polym. Sci.* **1988**, 266, 1087.

51.   He, Jing, *Crystallization Kinetics of Polymers at High Pressure*, Ph.D. Thesis, University of Colorado, Boulder, 1992.

52.   Mandell, J.F.; Smith, K. L.; Huang, D. D., *Polym. Eng. Sci.* **1981**, 21, 1173.

53.   Isayev, A. I.; Crouthamel, D. L., *Polym.-Plast. Technol. Eng.* **1984**, 22 (2), 177.

54.   Lafleur P., *Ph.D. thesis*, McGill University, 1983.

55.   Gatewood, B. E., *Thermal Stresses*, McGraw-Hill, N.Y., 1959.

56.   Titomanlio, G.; Brucato, V.; Kamal, M. R., *Int. Polym. Processing 1* **1987**, 2,55.

57.   Brucato, V. M. B.; Piccarolo, S.; Titomanlio, G., *Mat. Eng.* **1989**, 2 (2), 597-604.

58. Menges, G.; Dierkes, A.; Schmidt, L.; Winkel E., *Ibid.* **1980**, 26 , 300-306.
59. Douven, L., *PhD. Thesis*, Technische Universiteit Eindhoven, Eindhoven, Netherlands, 1991.
60. Cohen, A., *Modeling of Polymer Processing*, Ed. Isayev, A. I., Hanser Publishers, 1991.
61 Farhoudi, Y.; Kamal, M. R., *ANTEC Proceedings* **1994**, Vol. II, 1834-1837.
62 Farhoudi, Y.; Kamal, M. R., *44th. Canadian Chemical Engineering Conference*, Calgary, 1994.

RECEIVED March 14, 1995

Chapter 7

Advanced Modeling of the Generation and Movement of Gases Within a Decomposing Polymer Composite

Hugh L. N. McManus and David S. Tai

Technology Laboratory for Advanced Composites, Department of Aeronautics and Astronautics, Massachusetts Institute of Technology, Cambridge, MA 02139

The release of absorbed water has been shown to be a critical factor in the failure of composite material ablators. A reaction rate equation to model the phase change of water to steam in composite materials is derived from the theories of molecular diffusion and equilibrium moisture concentration. The model is dependent on internal pressure, the microstructure of the voids and channels in the composite materials and the diffusion properties of the material. Hence, it is more fundamental and accurate than the empirical equations currently in use. The model and its implementation into the thermostructural analysis code CHAR are described. Results of parametric studies on the variation of several parameters are presented.

When composite nozzle liner materials (geometry shown in Figure 1) are exposed to the high temperature environment inside a rocket nozzle, different reaction zones develop as shown in Figure 2. First, the material on the heated side will begin to decompose and form a pyrolysis zone. When the pyrolysis is complete, it leaves a layer of char behind. As more heat conducts into the material, the pyrolysis zone advances deeper into the virgin material. Ahead of the pyrolysis zone, absorbed moisture is released. A moisture evaporation zone will also develop and advance ahead of the pyrolysis zone in lower temperature material. Gases, which are produced by pyrolysis decomposition and moisture evaporation, flow through the material to the surface. These gases can cause large internal pressures.

The thickness of the composite insulation is designed so that the char layer will not reach the back side of the material before the rocket engine is shut down. However, several anomalous events can occur during the flight which can cause the insulation to fail prematurely. One of the severe anomalies is known as *ply-lift*. Ply-lift refers to the across-ply failure of the matrix material, and it has been observed in the exit cone liners of post-fired rocket engines. The ply-lift failure mode usually occurs in composites with low ply angles in the region just underneath the pyrolysis zone, as shown in Figure 3. Ply lift failure is discussed from an experimental point of view by Stokes (1).

The ply-lift failure has been attributed to the following mechanisms. When the material is heated rapidly, gases are generated and trapped. These gases cause a large increase in internal pressure which forces the plies apart. Since ply-lift usually occurs

0097–6156/95/0603–0114$12.00/0

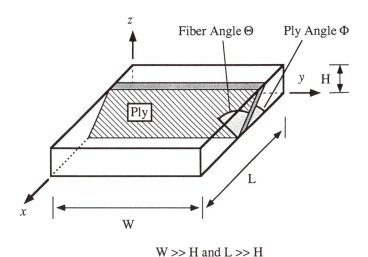

W >> H and L >> H

Figure 1. The geometry of ablative insulation.

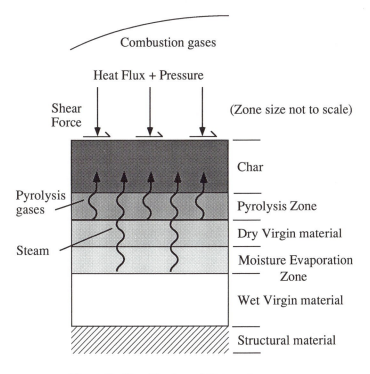

Figure 2. Classification of the reaction zones.

in carbon-phenolic composite materials at temperatures below 400 °C and pyrolysis reactions usually do not begin below 400 °C, it is suspected that the high pressure is mainly caused by the release of absorbed water (1). When the matrix material in the composite decomposes to char, the material's permeability can increase by as much as seven orders of magnitude. Thus, the gases generated inside the pyrolysis zone can escape easily, while steam released in the evaporation zone has more difficulty escaping since it has to pass through the relatively impermeable virgin material between the pyrolysis and moisture evaporation zones. Large internal pressures are built up by gases trapped between these zones. Although not yet pyrolyzing, the material in this region is hot enough to have lost much of its room temperature strength, and it is in this narrow region where ply-lift failure usually occurs. Better modeling of the moisture evaporation process will thus result in more accurate prediction of the internal pressure and ply-lift failure.

In this study, a new moisture release model, based on fundamental physical models of moisture diffusion and equilibrium moisture concentration, is derived. This model is coupled into an existing thermo-chemical-structural analysis program to provide a new and more accurate tool for predicting the behavior of ablative composite materials.

Background.

The techniques used in analyzing decomposing polymer composites are reviewed briefly in the introductory chapter to this book. Here, a specific review of the modeling of decomposition and moisture release in polymer matrix composites, as used in the overall modeling of thermo-mechanical analysis of decomposing structures, will be given.

In general, a chemical reaction rate can be modeled by an n-th order Arrhenius rate equation. Many authors have reported on rate equations for the decomposition of phenolic materials typically used in ablative applications, of which the most relevant is that of Henderson et al. (2). More current models for the chemical pyrolysis and gas generation of carbon-phenolic materials include the 4-step Arrhenius model used by Sullivan (3), and Kuhlman (4), originally developed by Loomis, et al. (5). The Arrhenius reaction is not pressure dependent. Therefore, it can predict the generation of gases even if the pressure at the point of generation is higher than the saturation pressure of the assumed gaseous substance. This is physically unreasonable. To model a temperature and pressure dependent moisture evaporation rate equation, McManus (6) proposed using a 2-step reaction including a simple straight-line model for moisture release, in which the reaction rate is constant. He later proposed a hybrid model with straight-line moisture release and Arrhenius pyrolysis, and also an Arrhenius equation with the activation energy as a function of pressure (7). Arrhenius rate reactions coupled with models for re-condensing of pressurized water have also been proposed. However, all these methods are empirical. There is no guarantee that they give an accurate moisture evaporation rate and they provide no insight into rate determining mechanisms.

Approach.

The new reaction rate model will consider the physical processes of molecular diffusion of moisture to the surface of a pore channel, and the release of moisture (steam) into the channel. For a given initial moisture content, diffusivity constants, and pore channel geometry, the release rate of moisture to steam as a function of temperature and time will be predicted. The new reaction rate equation will be incorporated into the existing CHAR code. Coupling the new reaction rate equation to the thermal, mass continuity and stress equations in CHAR will allow determination of pressure, stress and failure (if any) to be predicted as functions of time and position throughout the ablative structure.

Theory.

The new reaction rate equation for moisture evaporation will be derived from a microscopic point of view. First, it is assumed moisture is uniformly distributed within the material. Then the moisture must diffuse to a nearby pore channel, driven by a difference in concentration. The pore channel is either a long crack along the fiber-matrix interface or a closed crenular channel inside a fiber. The geometry considered is shown in Figure 4. A coordinate r is defined perpendicular to the fiber direction. Radial symmetry is assumed. The fiber direction is related to the laminate coordinate system by angles Θ and Φ (Figure 1). The moisture will evaporate at the pore channel's surface and become steam. How much moisture evaporates at the surface is determined by the equilibrium moisture concentration, which depends on the temperature and vapor pressure inside the pore channel. Since moisture evaporation on the surface at high temperature is very fast, it is assumed the surface will achieve equilibrium instantly. Then steam will escape by flowing through the pore channel, driven by a pressure difference. A schematic of the moisture release process is shown in Figure 5.

Moisture Diffusion. The moisture diffusion is governed by Fick's diffusion law (*8*).

$$\nabla \cdot (d\nabla c) = \frac{dc}{dt} \tag{1}$$

where d is the moisture diffusivity, c is the moisture concentration (the mass of moisture inside the material divided by the mass of dry material), and ∇ is the del operator. The diffusivity d is given by

$$d = d_0 \exp(-\frac{E_w}{R_g \cdot T}) \tag{2}$$

where d_0 is a pre-exponential factor, E_w is an activation energy, R_g is the universal gas constant, and T is the absolute temperature.

The moisture control volume is defined such that all moisture inside the control volume will go to one pore channel. In general, the boundary of a moisture control volume is an irregular polygon. This shape is approximated by a circle with radius r_a. The initial moisture content is assumed to be uniform inside the material, smearing any difference in moisture absorption between the fiber and the matrix. Similarly, the fiber and matrix materials around the pore channel are assumed to have a homogeneous effective diffusivity constant. If it is assumed that r_a is less than Δz, where Δz is the node spacing for numerical calculations, then it can also be assumed the temperature inside any one control volume is constant so that moisture diffusivity is constant everywhere in the control volume. With the same assumption, the derivative of c in the fiber direction can be neglected, and the moisture release rate can be derived independent of Θ and Φ. These assumptions reduce the problem from three dimensions to one, and equation 1 is reduced to

$$\frac{1}{r}\frac{\partial}{\partial r}(r\frac{\partial c}{\partial r}) = \frac{\partial^2 c}{\partial r^2} + \frac{1}{r}\frac{\partial c}{\partial r} = \frac{1}{d}\frac{\partial c}{\partial t} \tag{3}$$

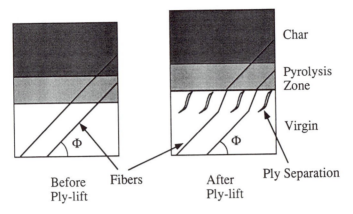

Figure 3. Geometry of ply lift failure.

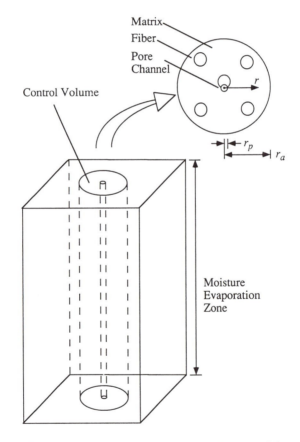

Figure 4. Geometry of the moisture release model.

An initial condition $c = c_0$ and boundary conditions

$$c = 0 \qquad \text{at } r = r_p$$

$$\frac{dc}{dr} = 0 \qquad \text{at } r = r_a$$

are assumed, i.e. moisture is zero at the surface of the central pore channel and moisture cannot cross the boundary to other control volumes. Using separation of variables, let

$$\frac{c(r,t)}{c_0} = R(r)\theta(t) \tag{4}$$

Then

$$\frac{R''}{R} + \frac{1}{r}\frac{R'}{R} = \frac{1}{\theta}\frac{\dot{\theta}}{d\,\theta} = -\lambda^2, \quad \lambda \geq 0 \tag{5}$$

where R' indicates the derivative of R with respect to r and $\dot{\theta}$ indicates the time derivative of θ. For the $R(r)$ part,

$$r^2 R'' + r R' + \lambda^2 r^2 R = 0 \tag{6}$$

This ordinary differential equation (ODE) has the form of Bessel differential function of order zero (9), and the general solution is

$$R(r) = A J_0(\lambda r) + B Y_0(\lambda r) \tag{7}$$

where A and B are constants, J_0 is a Bessel function of order 0 of the first kind and Y_0 is a Bessel function of order 0 of the second kind. The boundary condition at $r = r_p$ requires $R(r_p) = 0$, and the boundary condition at $r = r_a$ requires $R'(r_a) = 0$. These boundary conditions can only be satisfied if the following characteristic equation is satisfied. Details may be found in (10).

$$J_0(\lambda_n r_p) Y_1(\lambda_n r_a) - Y_0(\lambda_n r_p) J_1(\lambda_n r_a) = 0 \tag{8}$$

To simplify this characteristic equation, let

$$\lambda_n r_p = z_n \qquad\qquad \lambda_n r_a = \alpha z_n \tag{9}$$

where

$$\alpha = r_a / r_p \tag{10}$$

Then

$$J_0(z_n) Y_1(\alpha z_n) - Y_0(z_n) J_1(\alpha z_n) = 0 \tag{11}$$

The roots z_n are solved from equation 11 numerically for any given α, and λ_n is calculated by equation 9. Then equation 7 becomes

$$R(r) = \sum_{n=1}^{\infty} C_n y_n(r) \tag{12}$$

where

$$y_n(r) = Y_0(\lambda_n r_p) J_0(\lambda_n r) - J_0(\lambda_n r_p) Y_0(\lambda_n r) \tag{13}$$

$R(r)$ is represented by the sum of the magnitude C_n times the mode shapes $y_n(r)$.

For the $\theta(t)$ part,

$$\dot{\theta} = -\lambda_n^2 d\theta \tag{14}$$

Substituting equation 2 into equation 14

$$\dot{\theta} = -\lambda_n^2 d_0 \exp(-\frac{E_w}{R_g T})\theta \tag{15}$$

which is an ODE with solution

$$\theta(t) = \theta(0) \exp(-d_0 \lambda_n^2 g(t)) \tag{16}$$

where

$$g(t) = \int_0^t \exp(-\frac{E_w}{R_g T(s)}) ds \tag{17}$$

Since the initial conditions is $c(0) = c_0$, and $R(r)$ is assumed to be 1 for $r_p \le r \le r_a$ at time zero, $\theta(0) = 1$ and

$$\theta(t) = \exp(-d_0 \lambda_n^2 g(t)) \tag{18}$$

Applying $R(r)=1$ for $r_p \le r \le r_a$ to equation 12,

$$1 = \sum_{n=1}^{\infty} C_n y_n(r) \qquad r_p \le r \le r_a \tag{19}$$

Since $y_n(r)$ are orthogonal functions from the Sturm-Liouville Theorem (9), the constants C_n can be found by

$$C_n = \int_{r_p}^{r_a} r y_n(r) dr \bigg/ \int_{r_p}^{r_a} r y_n^2(r) dr \tag{20}$$

Therefore, the complete solution is

$$\frac{c(r,t)}{c_0} = \sum_{n=1}^{\infty} C_n y_n \exp(-d_0 \lambda_n^2 g(t)) \tag{21}$$

Let s be the total moisture concentration per unit length in the control volume, and apply $\alpha = r_a / r_p$ from equation 10,

$$s = c_0 \sum_{n=1}^{\infty} C_n \exp(-d_0 \lambda_n^2 g(t)) \int_{r_p}^{\alpha r_p} y_n(r) 2\pi r \, dr \tag{22}$$

The initial total moisture concentration per unit length in the control volume is

$$s_0 = c_0 \pi (\alpha^2 r_p^2 - r_p^2) = c_0 \pi (\alpha^2 - 1) r_p^2 \tag{23}$$

The degree of conversion C_w equals the mass of moisture inside the control volume divided by the initial mass of moisture inside the control volume. So

$$C_w = s/s_0 \tag{24}$$

Then the derivative of C_w with time equals the moisture release rate to the central pore channel. Let $u = r/r_p$ and transform the integral

$$\int_{r_p}^{\alpha r_p} y_n(r) \cdot 2\pi r \, dr = 2\pi r_p^2 \int_1^{\alpha} y_n(u) u \, du \tag{25}$$

where

$$y_n(u) = Y_0(z_n) J_0(z_n u) - J_0(z_n) Y_0(z_n u) \tag{26}$$

Finally, combining equations 20 through 26, the degree of conversion for zero moisture at the surface is

$$C_u = \sum_{n=1}^{\infty} D_n \exp(-d_0 \lambda_n^2 g(t)) \tag{27}$$

where

$$D_n = \frac{2}{\alpha^2 - 1} C_n \int_1^{\alpha} y_n(u) u \, du = \frac{2}{\alpha^2 - 1} \frac{(\int_1^{\alpha} y_n(u) u \, du)^2}{\int_1^{\alpha} y_n^2(u) u \, du} \tag{28}$$

and $g(t)$ is taken from equation 17.

Various values of z_n and D_n vs. α were calculated by Maple V (*11*), a general purpose mathematical software tool, and are shown in Tables I and II. The radius ratio α can be found by the following approximation

$$\alpha = \sqrt{1/n r_p^2} \tag{29}$$

where n is the number of pore channels per unit area.

Table I. z_n vs. α

α	z_1	z_2	z_3	z_4	z_5
2	1.36078	4.64590	7.81416	10.9671	14.1151
5	0.28236	1.13921	1.93918	2.7312	3.5204
10	0.11027	0.49788	0.85543	1.2087	1.5603
20	0.04651	0.23175	0.40160	0.5693	0.7362
50	0.01579	0.08793	0.15381	0.2189	0.2837
100	0.00717	0.04290	0.07546	0.1077	0.1397

Table II. D_n vs. α

α	D_1	D_2	D_3	D_4	D_5
2	0.87022	0.06299	0.02199	0.01113	0.00671
5	0.93079	0.03506	0.01150	0.00570	0.00341
10	0.95833	0.02220	0.00678	0.00327	0.00192
20	0.97438	0.01449	0.00411	0.00190	0.00109
50	0.98554	0.00881	0.00227	0.00099	0.00055
100	0.99002	0.00637	0.00155	0.00065	0.00035

As seen in Table I, higher mode shapes have larger z_n and hence, by equation 9, larger λ_n, and so they decay faster. As α increases, the z_n's decrease because more time is needed to diffuse a larger volume of moisture into the same pore channel. Table II shows that D_1 is much larger than other D_n's, and the magnitude drops off quickly for higher modes. If the radius ratio α is large, the series solution using only one term will give good results. Five terms of equation 27 are used here, which should give excellent results under realistic conditions.

Surface Equilibrium Moisture Concentration. In general, the boundary condition at $r = r_p$ will not be zero. The equilibrium moisture concentration is given by the following empirical equation (8,12),

$$c_\infty = c_{max}\left(\frac{P_v}{P_{sat}(T)}\right)^b \tag{30}$$

where P_v is the vapor pressure of water, P_{sat} is the saturated pressure of water, and c_{max} is the maximum moisture content. If one assumes that the moisture evaporation rate at the surface is very fast, then equilibrium is achieved at once at the surface. The boundary condition is then

$$c(r_p) = c_\infty \tag{31}$$

Now b is set equal to 1 and c_{max} equal to c_0. Even though supersaturated steam may exist inside the pore channel (i.e. $P_v > P_{sat}(T)$), c_∞ cannot be greater than c_{max} since the material could not physically absorb more moisture than the maximum amount. In that case, the supersaturated steam will probably condense to water inside the pore channels. This possible phenomena is neglected here so that a single phase flow of gas can be used.

Moisture Evaporation Rate. The solution for moisture diffusion and concentration derived in the previous section assumed a boundary condition of zero concentration at the surface. The solution for realistic boundary conditions can be found by a convolution integral (*10*). The degree of conversion is given by

$$
\begin{aligned}
C_w(t) &= f(t) - \int_0^t C_u(t-u) \frac{df}{dt} du \\
&= f(t) - \int_0^t \dot{f}(u) \sum_{n=1}^{\infty} D_n \exp(-\lambda_n^2 d_0 [g(t) - g(u)]) du
\end{aligned}
\tag{32}
$$

where equilibrium degree of conversion $f(t) = c_\infty / c_0$ is given by

$$
\begin{aligned}
f(t) &= \frac{c_{max}}{c_0} \left(\frac{P_v}{P_{sat}}\right)^b \qquad P_v < P_{sat} \\
f(t) &= \frac{c_{max}}{c_0} \qquad\qquad P_v \geq P_{sat}
\end{aligned}
\tag{33}
$$

and $f(t)$ is always constrained to be less than or equal to one. The beta of moisture or degree of dry-out is

$$
\beta_w = 1 - C_w
\tag{34}
$$

And the desired steam generation rate is

$$
G_w = -c_0 \rho_s \frac{dC_w}{dt} = c_0 \rho_s \frac{d\beta}{dt}
\tag{35}
$$

Numerical Method. The steam generation rate (equation 35) is calculated numerically by a routine that is embedded in the CHAR code. Temperature and pressure conditions are provided by CHAR at each time step. These determine the boundary conditions. The degree of conversion C_w is found by numerically integrating equation 32 using the same time steps as the rest of the CHAR solution, and the change in degree of conversion from the previous time step provides the generation rate. Details of the numerical method are provided in (*10*).

Parametric Study and Results.

The new moisture release rate model has been incorporated into the CHAR computer code. A standard case was established as a baseline for parametric studies. Based on observed geometry (*1*), r_a equals 1 μm and r_p equals 20 μm, and diffusion constants were taken from (*13*). A CHAR model with 1001 nodes and 0.5 sec maximum time step was used to perform the studies. Despite the large number of nodes, typical run time on an IBM RS6000 320H workstation was 4 to 5 minutes. Parametric studies of various pore sizes, pore spacing, and diffusivity constants were performed.

The baseline CHAR analysis case used in previous studies (*7*) was used here. A plate with 3 cm height, 45 degree Θ, 15 degree Φ, and initial moisture content 3.5% was analyzed. The boundary conditions used represented a simplified rocket nozzle service environment. The external temperature and pressure will rise to 3000 K and 10 MPa respectively after the rocket ignites. At 100 sec, the rocket motor is shut off and the external temperature and pressure are assumed to ramp down in 5 sec to 300 K and 0.1 MPa respectively. CHAR outputs the temperature, internal pressure, mass flow, degree of char, degree of dry-out, and stresses. The maximum pressure

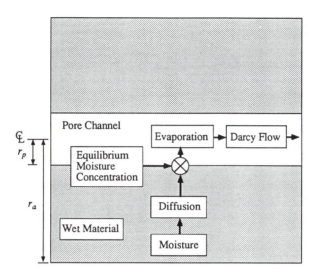

Figure 5. Schematic of the moisture release model.

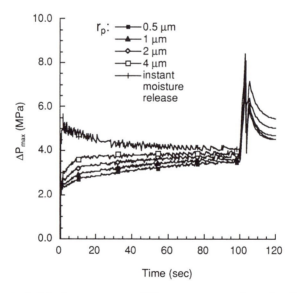

Figure 6. Maximum pressure difference vs. pore channel radius.

difference (internal pressure minus external pressure), which is the driving force behind ply lift, will be used for comparison in all of the parametric cases studied here. At shut off, the external pressure drops suddenly, causing an increase in the pressure difference. This appears as a spike in many of the figures in this section. The results from the baseline case numerically replicate a ply-lift failure after motor shut off due to this spike. However, if the internal pressure rises faster than in this example case, ply-lift failure can occur before shut off, which could cause premature nozzle failure. Increased pressure can be caused by variation of many parameters: lower permeability, smaller angle Φ, higher initial moisture content, and lower external pressure are examples (7).

Effects of Pore Size. According to Stokes (1), the diameter of closed crenular channels is 1 to 3 μm. Therefore, r_p is set equal to 4, 2, 1, and 0.5 μm and the maximum pressure difference histories are plotted in Figure 6. The maximum pressure difference is slightly higher for larger pore sizes. Figure 6 also shows a curve for instant moisture release, which illustrates the limit at which diffusion is instantaneous.

Effects of Pore Spacing. The control volume radius r_a was set equal to 5, 10, 20, 40, and 100 μm. As r_a increases, α increases such that all z_n decrease (see Table I) and the moisture release takes a longer time to finish. As shown in Figure 7, the maximum pressure difference is notably lower for larger r_a, while for small r_a the curve approaches the instantaneous release limit shown in Figure 6.

Figures 8 and 9 show the propagation of pyrolysis and moisture evaporation zones for the r_a equals 20 μm and r_a equals 40 μm cases respectively. The zones are defined as the regions where the values of pyrolysis or moisture degrees of conversion fall between 5% and 95%. The edges of the pyrolysis zone are shown by solid lines, and the edges of the evaporation zone by dotted lines. Figure 8 shows the moisture evaporation zone proceeding ahead of the pyrolysis zone. After shut off, external pressure drops to 0.1 MPa and the moisture evaporation zone expands because more moisture is able to evaporate at lower pressure. The additional convective cooling from this moisture tends to stop the progress of the pyrolysis reaction. Figure 9 shows the effect of larger pore spacing. Moisture release is delayed, and the moisture release and pyrolysis zones merge. Because the permeability inside the pyrolysis zone is much greater than that of the virgin material, steam generated inside the pyrolysis zone can escape easily and causes a much smaller pressure rise. Little moisture is released ahead of the pyrolysis zone, resulting in decreased pressure difference, at least until shut off.

Effects of Diffusivity Constants. The pre-exponential factor d_0 cannot be determined very accurately. For the same material, d_0 from different measurements can differ by as much as 2 orders of magnitude (12). Therefore, the nominal d_0 is varied by factors of ten times larger and smaller and the maximum pressure difference vs. time are plotted in Figure 10. The effect is quite noticeable, as one would expect given the drastic changes in d_0.

For E_w/R_g, the range of available data is about 500 K (13). Therefore, E_w/R_o is varied by +500 K and -500 K. The corresponding maximum pressure differences are plotted in Figure 11. A relatively small effect on the maximum pressure difference prediction is seen.

Comparisons to Other Rate Models. The new model was compared to the straight-line and Arrhenius models for moisture release. The original straight-line model assumed that moisture was released uniformly over a temperature range ΔT of 50 K. A ΔT of 100 K was also considered. The Arrhenius model was a 4-part combined

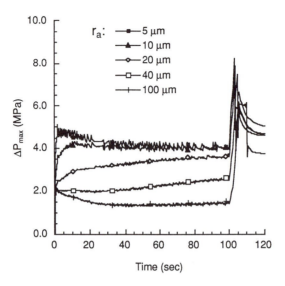

Figure 7. Maximum pressure difference vs. r_a.

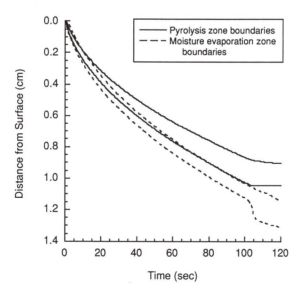

Figure 8. Reaction zone history for $r_p = 1$ μm and $r_a = 20$ μm.

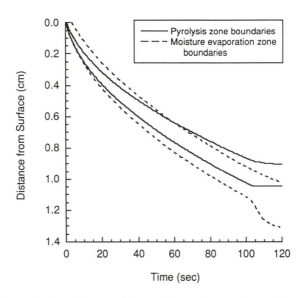

Figure 9. Reaction zone history for $r_p = 1$ μm and $r_a = 40$ μm.

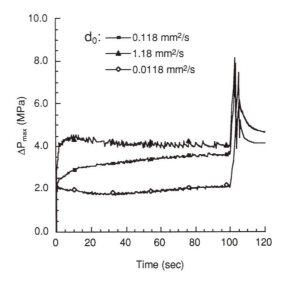

Figure 10. Maximum pressure difference vs. diffusion pre-exponent factor.

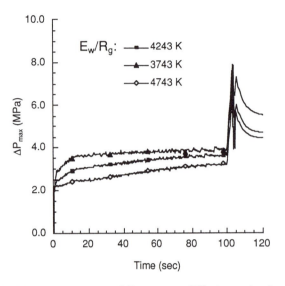

Figure 11. Maximum pressure difference vs. diffusion activation energy.

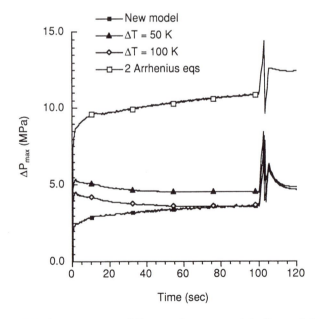

Figure 12. Maximum pressure difference for new, straight-line and Arrhenius models.

moisture release and pyrolysis model (7). Only the first two parts, which are assumed to involve moisture release, were used. The pyrolysis model used was the same in all cases.

Figure 12 shows the maximum pressure differences calculated for the new, straight-line and Arrhenius models. The Arrhenius model overpredicts the maximum pressure difference because it is not pressure dependent. The straight-line model overpredicts the maximum pressure difference early in the simulation because it is not time dependent. A straight-line model, especially with $\Delta T = 100$ K, gives good agreement at low temperature rise rates typical in the later part of the simulation. At high heating rates prevalent in the early part of the simulation, the straight-line model predicts faster moisture release, resulting in a higher maximum pressure difference.

Conclusions.

A new method for calculating moisture release rates is based on a micro-scale model of moisture diffusion to a nearby pore channel and moisture evaporation on the pore channel surface. The diffusion of moisture causes the moisture release rate to be both time and temperature dependent. The equilibrium condition on the pore channel surface causes the moisture release rate to be dependent on pressure. The method has been expressed mathematically and implemented as a module of the CHAR code. It was found that only a few terms of the series solution were necessary for accurate results, resulting in good computational efficiency.

The inclusion of diffusion in the model slows the predicted release of moisture. This causes moisture to be released at higher temperatures. This tends to remove the separation between the moisture evaporation and char zones, and results in lower predicted pressures.

The geometry of the pores strongly affects the moisture release rate. Pore spacing has a larger effect than the pore size. Larger spacing (or smaller pore size) slows diffusion to the pores and results in lower predicted pressures. Very large pore spacing slows the diffusion so much that the effect of moisture on predicted pressure is almost lost. Very small pore spacing allows very rapid moisture release to the limit that the diffusion effect is lost.

The value of the diffusivity rate constants have a lesser effect on the moisture release rate. Varying the values of the activation energy E_w well outside the measured range had only a moderate effect on the moisture release and pressure. Varying the rate constant d_0 by two orders of magnitudes changed the calculated pressure difference by a factor of two.

Comparison of the new model with existing Arrhenius and straight-line models illustrates how the more fundamental nature of the new model produces more realistic results. The new model is time, temperature, and pressure dependent, and so it produces physically realistic results under all conditions.

Literature Cited

1   Stokes, E.H. In *Computational Mechanics of Porous Materials and Their Thermal Decomposition*; Salamon, N. J., and Sullivan, R. M., Eds.; AMD-136, American Society of Mechanical Engineers: N. Y., NY, 1992, pp 146-156.
2   Henderson, J. B.; Wiebelt, J. A.; Tant, M. R. *J. of Composite Materials* **1985,** *19,* 579-595.
3   Sullivan, R. M.; Salamon, N. J. *Int. J of Engineering Science,* **1992,** *30,* 431-441.
4   Kuhlmann, T. L. *Thermo-Chemical-Structural Analysis of Carbon-Phenolic Composites with Pore Pressure and Pyrolysis Effects;* Ph.D. Thesis; University of California: Davis, CA, 1992.

5   Loomis, W. C.; Clayton, F. I.; Hieghtland, C. N., *Process Chemistry Definition for 2D Carbon-Carbon Fabrication for Solid Rocket Motor Technology*; Final Report under Contract NAS8-36294; National Aeronautics and Space Administration: George C. Marshall Space Flight Center, AL, 1988.
6   McManus, H. L.; Springer, G. S. *J. of Composite Materials* **1992,** *26,* 206-229.
7   McManus, H. L. N. In *Computational Mechanics of Porous Materials and Their Thermal Decomposition*; Salamon, N. J., and Sullivan, R. M., Eds.; AMD-136, American Society of Mechanical Engineers: N. Y., NY, 1992, pp 113-119.
8   Tsai, S. W.; Hahn, T. H. *Introduction to Composite Materials*; Technomic Publishing Company: Lancaster, PA, 1980.
9   Wylie, C. R.; Barret, L. C. *Advance Engineering Mathematics, 5th edition;* McGraw-Hill: N.Y., NY, 1982.
10 Tai, D. S. *A New Model of Moisture Evaporation in Composite Materials in Rapid Temperature Rise Environments*; Master of Science Thesis; Department of Aeronautics and Astronautics, Massachusetts Institute of Technology: Cambridge, MA, 1993.
11 Maple V, Waterloo Maple Software, University of Waterloo, Ontario, Cananda.
12 Sih, G. C.; Michopoulos, J. G.; Chou, S. C.; *Hygrothermoelasticity*; Martinus Nijhoff Publishers: Boston, MA, 1986.
13 Stokes, E.H. In *Proceedings of the 1987 JANNAF Rocket Nozzle Technology Subcommittee Meeting*; Chemical Propulsion Information Agency: Columbia, MD, 1987.

RECEIVED February 22, 1995

Chapter 8

The Application of Supercritical Fluid Technology to High-Performance Polymers

J. R. Lee[1] and R. G. Kander[2]

[1]Materials Engineering Science Program and [2]Materials Science and Engineering, Virginia Polytechnic Institute and State University, 213 Holden Hall, Blacksburg, VA 24061

The application of supercritical fluid (SCF) technology to high-performance polymers is becoming increasingly important as such polymers become more prevalent in industry. How fabricated parts respond to SCF environments can be critical to performance. The technology of powder precipitation of these polymers from SCF solutions is slowed by the lack of a body of knowledge regarding the solubility of main-chain aromatic polymers. Recent studies of aliphatic polymers with supercritical fluid solvents and of aromatic polymers with liquid solvents serves are reviewed. These studies and several presented applications draw attention to some industrial needs and research opportunities in the area of high-performance polymer/supercritical fluid solvent interaction.

Research on the dissolution of small molecules and aliphatic polymers in supercritical fluid (SCF) solvents is commonplace in the literature (1,2). Recent interest has been expressed in dissolving high-performance polymers in SCFs. In certain applications SCFs are attractive over liquid solvents because of their lower density, higher thermal diffusivity and higher mass transport. These properties may be useful in the elimination of potentially dangerous liquid solvents in polymer processing. These dangerous aspects include flammability, toxicity and carcinogenicity. Additionally, SCFs may offer post-processing advantages such as a lack of residual solvents. This may serve to enhanced structural integrity through void volume reduction.

A review of recent literature reveals a scarcity of reports which apply SCF technology to main-chain aromatic polymer systems, or those containing benzene rings along the backbone. This paper will detail some of the current applications of SCF technology to the polymer field and cover some of the limitations encountered inapplying these techniques to high-performance polymer systems. Some areas of recent work regarding the utility of supercritical fluids are then addressed.

0097–6156/95/0603–0131$12.00/0

Synthesis

A well-recognized application of SCFs in polymer synthesis is the production of low density polyethylene (LDPE) from supercritical ethylene. This *high-pressure* technique was developed by ICI in the 1930's (*3*). This process typically operates between 1500 and 3000 atmospheres, while temperatures are kept below 300°C to avoid polymer decomposition. Molecular weight, molecular weight distribution and the degree of chain branching are well-controlled functions of the SCF conditions as well as the choice of initiator and chain-transfer agent levels.

Some researchers have turned to replacing a certain percentage of hydrogen with fluorine to improve high-temperature resistance. This makes the chain bulkier, generally reducing the tendency to crystallize and increasing solubility (*4*). Research involving the fluorination of acrylic monomers has successfully yielded aliphatic homopolymers soluble to 25%(w/v) in supercritical carbon dioxide (*5*). These researchers have also investigated the copolymerization of these fluorinated monomers with olefinic monomers, as homopolymers of these olefinic groups have been determined to be insoluble in CO_2. However, the fluorinated monomer is soluble enough to incorporate over 55% of the olefinic group into the copolymer (*6*).

No main-chain aromatic polymerizations in SCFs are noted in the literature. The synthesis of aromatic polymers in supercritical fluids is more difficult because of the inherent insolubility of such large monomers and the stiffness of the polymerizing chain.

Swelling

The absorption of SCFs into polymers has been of commercial interest. Applied research includes the effect of gases on elastomeric seals in applications such as oil recovery (*7*) or where these gases are otherwise under high pressure (*8*). Further work with the amorphous polymers--silicone rubber and polycarbonate--studied the uptake of carbon dioxide at pressures up to 60 atmospheres (*9*). These researchers determined that the interactions of the CO_2 with the high level of carbonyls in the polycarbonate enhanced swelling.

Goel and Beckman employed specific polymer/solvent interactions to produce microcellular foams (*10*). Poly(methyl methacrylate) (PMMA) was swelled with supercritical carbon dioxide. Subsequent vitrification by quick pressure reduction froze in the micropores, producing a foam. This process is referred to as *pressure induced phase separation* (PIPS).

A similar concept has been applied to liquid crystalline foam formation (*11*). Researchers used CO_2 at near-critical conditions to produce high-performance isotropic foams. These foams have potential application in the automotive and aerospace industries, where the combination of low density and high stiffness and toughness is desired.

High-performance polymer composites are being utilized for a variety of applications as replacements for metal parts. Many of these are subjected to SCF conditions during use.   For example, a manufacturer of supercritical fluid chromatography (SFC) equipment now offers a poly(aryletheretherketone) [PEEK] composite column (*12*). Also, a fiber-reinforced, PEEK composite called Kadel™ is being used to make small, durable, lightweight parts by an industrial compressor manufacturer (*13*).   These parts are used in the presence of halogenated solvents at elevated temperatures and pressures.   Manufacturers circumvent potential exposure problems and maintain structural integrity by incorporating fibers and by precrystallizing PEEK to a high level.

In many applications one must consider the durability of such parts in the presence of harsh solvents. Studies of the effect of such exposure on performance is necessary.   Testing conducted on neat, amorphous PEEK indicates that SCFs are soluble in PEEK and can cause solvent-induced crystallization (SINC) (*14*). A PEEK sample exposed for 24 hours to supercritical chlorodifluoromethane (R-22) swelled the polymer by approximately 10% (w/w). Figure 1 illustrates the clear, amorphous sample and the swollen, opaque sample. To prove the crystallizing effect that SCF R-22 has on PEEK, DSC scans were made on the finished sample with and without exposure (Figure 2). The absence of the recrystallization peak in the exposed sample is evidence of a crystallized sample. Also the small angle x-ray spectroscopy (SAXS) scan of the exposed sample reveals a long-spacing of about 10 nanometers (Figure 3).

Other types of polymers behave differently in the presence of SCFs. For example, samples of bisphenol-A polyphenylsulfone, an amorphous polymer, were subjected to the same SCF conditions as PEEK. The sample maintained its shape, but turned from completely clear to opaque. However, the opacity is due to foaming, a result similar to the foaming processes reported earlier. A fluorinated version of this sulfone (6F polysulfone) was subjected to the same conditions. The sample flowed and foamed also.

Recent polymer studies involving solvents include their effects on polymer aging (*15*), gas transport (*16*), swelling (*17,18*), and SINC (*19*).   However, none report the effects of SCF solvents on morphology or durability.   In addition to the

(a)   (b)

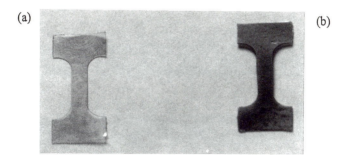

Figure 1.   Photographs of: (a)amorphous PEEK and (b)sample after 24-hr exposure to supercritical R-22.

interesting foaming effect illustrated above, other questions might be duly addressed regarding exposure. As many studies strive to determine the ability of liquid and SCF solvents to dramatically swell and solubilize polymers, one might also consider the long-term effects of relatively poor solvents. For example, this may be of particular interest to those seeking to use high-performance polymeric materials in traditional metal applications. Some questions to be addressed include: Over the long term, how is exposure to certain solvents going to possibly affect crystallinity? How might thermal expansion and durability be affected? Predicting the response of high-performance polymers to SCFexposure could be crucial to their continued expansion into roles traditional to commodity polymers.

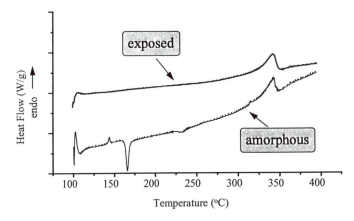

Figure 2. DSC scan revealing crystallinity caused by SCF-exposed PEEK.

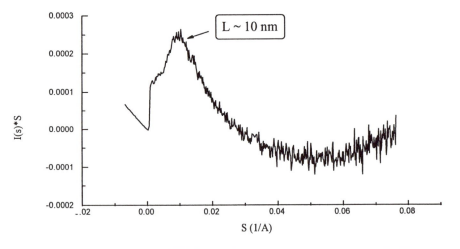

Figure 3. SAXS scan of PEEK after 24-hr exposure to supercritical R-22.

Solubility

Choosing an appropriate solvent for the dissolution of polymers at supercritical conditions often involves consideration of polarizability, hydrogen bonding potential and polarity matching. Goel and Beckman have studied the prerequisites for solubility in supercritical CO_2. They concluded that this solvent does not generally have an affinity for hydrocarbon polymers. However, as one increasingly incorporates inorganic species such as fluorine and silicon atoms, the polymer has a greater affinity for CO_2 (20).

McHugh et al have illustrated the dependence of solubility on group functionality in the copolymer poly(ethylene-*co*-methyl acrylate) (21). They found that by increasing the degree of the more polar acrylate functionality over ethylene, the polymer became more soluble in supercritical chlorodifluoromethane (R-22). Conversely, a systematic increase in ethylene functionality made the polymer more soluble in supercritical propane.

Recent work by the authors has involved the solubility of high-performance polymers in SCFs. Two aromatic polyphenylsulfones were investigated: an amorphous, bisphenol-A-type polysulfone analogous to Udel™ and the fluorinated version of the same, known as 6F polysulfone, synthesized at Virginia Tech by Dr. McGrath's group. These polymers had weight average molecular weights of 18,000 and 20,000, respectively. The solvent was chlorodifluoromethane, known commercially as Genetron™ 22 or generically as R-22.

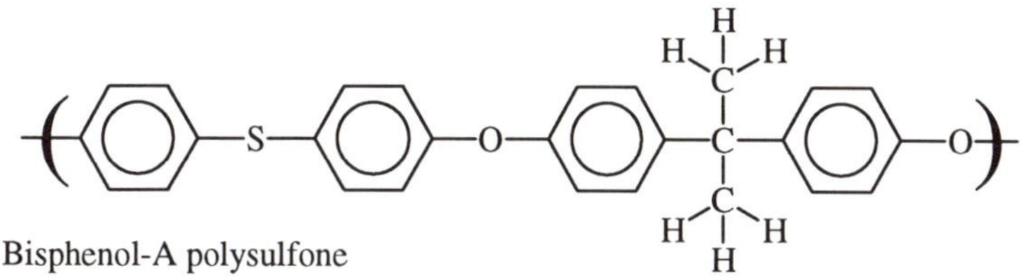

Bisphenol-A polysulfone

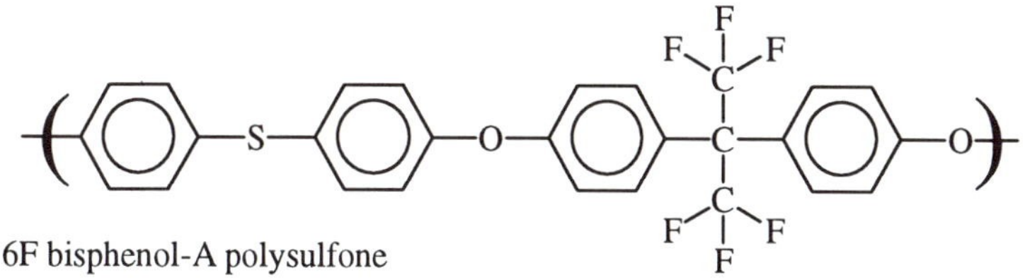

6F bisphenol-A polysulfone

Slight solubility of the 6F polysulfone was observed at pressures as low as 250 atmospheres and temperatures as low as 100°C (Figure 4). Solubility of the hydrogenous version was not observed. The solubility behavior was confirmed by the following reasons. Solubility was repeatable for a range of concentrations. There

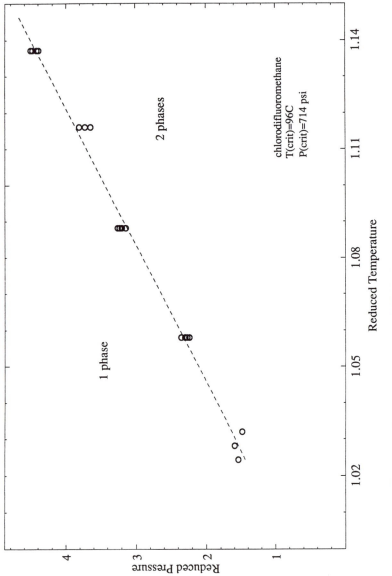

Figure 4. Cloud-point curves of 6F bisphenol-A polysulfone in supercritical chlorodifluoromethane.

was a definite lack of observed solubility in the hydrogenous moiety. The previously compared swelling behavior of both materials shows the fluorinated version has a significant affinity for the solvent over the hydrogenous version.

The observed solubility of the 6F polysulfone can be explained in terms of the dipole-dipole interactions between the acidic hydrogen of the solvent and the electronegative fluorinated methyl groups on the polymer. The expected hydrogen-bonding interaction between the same acidic hydrogen and the sulfone group of the polymer would not help to explain the lack of solubility of the non-fluorinated moiety.

Just as a broad range of methods are available to the polymer scientist interested in solubility in liquid solvents, future needs include the adaptation of such tools to SCF solubility. For example, the group contribution method to the solubility parameter concept could include the effect of the reduced densities of SCF solutions on coordination number for given systems. Also local composition equation of state models could better predict solubility behavior by considering the unique coordination numbers developed in SCF solutions.

Precipitation

Comminution of high-performance polymers into fine powders can allow a greater scope of processing opportunities. Of current interest is dry, solid-state processing of polymeric powders. The powder is cold compacted into a part with dimensional stability, after which it is sintered above the melting temperature (*22*). This process has been applied to ultra high molecular weight polyethylene (UHMWPE), where high viscosity and processing temperatures prohibit the application of conventional processing techniques (*23*).

An interesting technique utilizing neat polymer powders is called selective laser sintering (SLS) (*24*). A very thin layer of powder is laid on a surface (~0.010 inch). A laser is then programmed to scan the entire surface, turning on and off to selectively fuse together those portions of the layer that would become the finished part. Another layer of powder is laid on top of this processed layer and the laser again fuses the polymer. This continues until the finished part is composed of polymer powder which the laser ignored and a fused polymer part. Lastly, one may electrostatically prepreg fiber tow with these powders to produce prepreg tape for filament winding, as has been demonstrated with LaRC-TPI polyimide and with PEEK (*25,26*).

Polymer powders are currently produced by mechanical grinding, by air impingement, and by liquid solution precipitation. The precipitation of polymer powders from SCF solutions can possibly eliminate some of the problems associated with these methods such as broad particle-size distributions, the inefficiency of grinding low modulus polymers, and possible residual solvents.

Powder production via precipitation from SCF solutions presents economic and environmental advantages as a dry powder processing technique to replace current solvent precipitation methods because of the high thermal diffusivity of SCFs and potential lack of toxic solvent emissions. Polymer powder precipitation has been accomplished with a variety of hydrocarbon, halogenated organic, and carbon dioxide SCFs (27).

The rapid expansion of a supercritical fluid (RESS) beyond a nozzle or orifice has been used to precipitate polymer powders. The resulting morphology of the powders has been reported (28). Spheres and fibers have been produced. One can produce different particle shapes, depending on whether the mechanism of phase separation is by nucleation and growth or by spinodal decomposition.

Very little has been published regarding the precipitation of aromatic polymer powders (29). As solubility of these polymers is of course a prerequisite to precipitation, dissolution must first be addressed, as discussed in the previous section. SCFs may present an advantage to better solvate these polymers through the ability to vary the solvent density and their enhanced mass diffusion over liquid solvents.

Conclusion

Supercritical fluid technology has been deemed invaluable in a number of polymer applications from synthesis to solubilization to powder precipitation. The application of this knowledge to high performance polymers will come with a better understanding of polymer/solvent interactions at supercritical conditions and with the continued study of phase separation as a mechanism to precipitate various polymer powder morphologies.

Acknowledgment

We acknowledge William Marsillo and Andrea Hivner for their significant contributions in the lab with data collection and analysis. We acknowledge Dr. McGrath's group of the Department of Chemistry at Virginia Tech for providing polyphenylsulfone derivatives. We acknowledge the National Science Foundation Science and Technology Center for High Performance Polymer Adhesives and Composites at Virginia Tech under contract DMR #912004.

Literature Cited

1. Tsekhanskaya, Y. V.; Iomtec, M. B.; Mushkina, E. V. *Russ. J. Phys.* Chem.; **1964**, *38*, 1173.
2. Heller, J. P.; D. K. Dandge; R. J. Card; L. G. Donaruma. *Paper presented at the Int. Symp. on Oilfield and Geothermal Chem., Denver, CO,* **June 1983**.
3. *Olefin polymers: high pressure polyethylene* in Kirk-Othmer Encyclopedia of Chemical Technology, 3rd ed., John Wiley and Sons: New York, 1984, Vol. 16; 402.
4. Lyle, G.; Priddy, D.; McGrath, J. E. Department of Chemistry, Virginia Tech, unpublished data, 1994.
5. Clark, M. University of North Carolina--Chapel Hill, personal communication, 1994.
6. DeSimone, J. M.;Guan, Z.; Elsbernd, C. S. *Science* **1992**, *257*, 945.
7. Ender, D. H. *Chemtech* **1986**, 52.
8. Schroder, E.; Arndt, K. F. *Faserforsch. und Textiletech* **1976**, *27*, 135.

9.   Fleming, G. K.; Koros, W. J. *Macro.* **1986** *19*(8), 2285.
10.   Goel, S. K.; Beckman, E. J. *J. cell. polym.* **1993**, *12*(4), 251.
11.   Risch, B. G.; Wan, I.; McGrath, J. E.; Wilkes, G. L. accepted for publication, *J. of App. Poly. Sci.*, **1994.**
12.   Alltech, Inc., Deerfield, IL.
13.   Bristol Compressors, Product literature, Bristol, VA 24201.
14.   Lee, J. R.; Hivner, A.; Kander, R. G. Materials Science and Engineering Department, Virginia Tech, manuscript in preparation, **1995**.
15.   Michele, A.; Vittoria, V. *Polymer* **1993**, *34*(9), 1898.
16.   Michele, A.; Vittoria, V. *Poly. comm.*, **1991**, *32*(8), 232.
17.   Wolf, C. J.; Bornmann, J. A.; Grayson, M. A. *J. of poly. sci.: Part B: poly. phys.* **1991**, *29*, 1533.
18.   Stuart, B. H.; Williams, D. R. *Polymer* **1994**, *35*(6), 1326.
19.   Beckman, E.; Porter, R. S. *J. of poly. sci.: Part B: poly. phys.* **1987**, *25*, 1511.
20.   Goel, S. K.; Beckman, E. J. *J. of supercrit. fluids* **1992**, *5*, 237.
21.   Meilchen, M. A.; Hasch, B. M.; Lee, S.; McHugh, M. A. *Polymer* **1992**, *33*(9), 1922.
22.   Jog, J. P. *Adv. in poly. tech.* **1993**, *12*(3), 281.
23.   Han, K. S.; Wallace, J. F. *J. of macro. sci.--phys.* **1981**, *B19*(3), 313.
24.   Kimble, L. L. In *Solid freeform fabrication proceedings*, Markus, H. L., Ed. University of Texas at Austin, **Sept. 1992**.
25.   Baucom, R. M.; Marchello, J. M. *NASA, Langley Research Center*, Hampton VA 23665-5225, April, **1990**, 49 pp.
26.   Muzzy, J.; Varughese, B.; Thammongkol, V.; Tincher, W. *SAMPE Jour.* **1989**, *25*(5), 15.
27.   Matson, D. W.; Petersen, R. C.; Smith, R. C. *J. of Mat. Sci.* **1987**, *22*, 1919.
28.   Lele, A. K.; Shine, A. D. *AIChE Journal* **1992**, *38*(5), 742.
29.   Petersen, R. C.; Matson, D. W.; Smith, R. D. *Poly. eng. and sci.* **1987**, *27*(22), 1693.

RECEIVED March 14, 1995

Chapter 9

Degradation Kinetics of High-Performance Polymers and Their Composites

J. M. Kenny[1] and L. Torre[2]

[1]**Institute of Chemical Technologies, University of Perugia,
Loc. Pentima Bassa, 05100 Terni, Italy**
[2]**Department of Materials and Production Engineering, University
of Naples, P. Tecchio, 80125 Naples, Italy**

An experimental and theoretical study of the thermal stability of high performance thermoplastic matrices and their composites is presented. The study is focused on the degradation behavior, under different environments, of polyetheretherketone (PEEK), polyetherimide (PEI) and their continuous carbon fiber reinforced composites. The experimental results, obtained by thermogravimetric analysis performed in air and pure nitrogen, are the basis of the development of a phenomenology kinetic model which is able to predict the degradation rate under different environmental conditions. Moreover, the results of the model can be used to predict the onset of the degradation process at longer times and lower temperatures than those used in the experimental characterization. The experimental characterization has shown a good thermal stability of the materials studied with a better performance of PEEK with respect to PEI at high service temperatures.

Advanced thermoplastic polymers have been proposed in the last decade as a valid alternative to typical thermoset matrices in composites for aerospace applications. The most common thermoplastic matrices for these materials are the semicrystalline polyetheretherketone (PEEK) and the amorphous thermoplastic polyetherimide (PEI). Although cost considerations have not allowed an extensive utilization, thermoplastic matrix laminates normally offer higher toughness and damage tolerance properties than their equivalent thermoset based composites. As new projects in the aerospace industry are focused in the development of supersonic planes characterized by higher levels of the normal service temperatures, polymer matrix composites are being also considered as natural candidate materials as a consequence of their good thermal stability. This new application requires a reliable prediction of the long term behavior of such materials after exposure at elevated temperatures.

The utilization of a material under critical temperature conditions requires the knowledge of its degradation kinetics which is related both to morphological changes and to chemical decomposition reactions. The degradation of polymeric composites has been studied mainly in relation with the reprocessability of thermoplastics after several cycles (*1-3*). For example, PEEK changes irreversibly its ability to crystallize if it is repetitively heated up to 400°C. However, chemical degradation is typically detected by monitoring weight changes occurring to a polymer exposed to high

0097–6156/95/0603–0140$12.00/0

temperatures as a function of time. While in such situation metallic materials show generally a weight gain curve as a consequence of oxidation processes, polymers are characterized by weight loss processes that can be attributed to evaporation of solvents and/or plasticizers at relative low temperatures, and to chain scission with production of low molecular weight volatile molecules at higher temperatures. The particular mechanisms of degradation are a function of the polymer structure and can follow different steps because of the complexity of the polymeric chains (*4-6*). Functional group transfer, chain unzipping with elimination of volatiles, and the formation of intermediate compounds are the main decomposition mechanisms that contributes to complicate the development of a mechanistic kinetic model of the polymer degradation process. In particular, reported data on PEEK degradation indicated benzoquinone as the first gaseous product of the chain breaking reaction followed by the rearranging of the polymer chain (*6,7*).

The effect of the environment is very important in polymer degradation. The presence of oxygen as reactant combined with heat causes very rapid degradation processes in polymeric chains. Therefore its very important to fix the environmental condition, when decomposition kinetics is studied. Moreover, in the particular case of polymer based composites, the presence of the carbon fibers must also be considered as they can affect the degradation process and degrade themselves in the presence of oxygen. Furthermore, carbon fibers can affect the thermal diffusion in the bulk.

Thermogravimetric analysis (TGA) is commonly used to study the degradation of polymeric materials (*8*). With this technique the weight loss is measured as a function of time and temperature. Then, the degradation kinetics can be studied correlating the weight loss with a generic degree of degradation (*9,10*). Although TGA data may not give enough information about complex degradation mechanisms, it can be a very useful tool to analyze the overall behavior of the polymer in extreme conditions (*4, 11, 12*).

When long term data are needed, the main experimental problem is the test duration. However, accelerated degradation experiments at high temperatures can be performed and the long term behavior can be extrapolated at lower temperatures . This procedure can be applied, under particular restrictions, assuming a simplified degradation process (*9*). In this work the results of several TG tests, performed under different environmental conditions, are used to analyze the thermal stability, and to develop a kinetic model of their degradation processes. The model can be applied to extrapolate the degradation behavior of the materials studied at long times.

Experimental Procedures and Materials

Degradation studies were performed on pellets of polyetheretherketone (PEEK) produced by ICI and polyetherimide (PEI-Ultem1000) produced by General Electric, and on their commercial carbon fiber composites: APC2 and CETEX respectively. The materials used in this research were kindly provided by Alenia.

PEEK is a semicrystalline thermoplastic polymer that has been specially developed for high performance composites (*13*). It has a Tg of 147°C, a crystallinity content developed under normal processing conditions of 30%-35% with a melting point of 335°C and a processing temperature of 400°C. On the other hand PEI is an amorphous thermoplastic polymer with a Tg of 210°C and a processing temperature of 320°C. The environmental behavior of these two polymers has been mainly studied with reference of their solvent resistance (*14*).

Weight loss determinations were performed in a TA TGA Thermogravimetric Analyzer model 911. Samples were tested both in air and in nitrogen atmospheres, in isothermal and dynamic modes, at different temperatures and heating rates in order to cover a wide range of thermal conditions. Data obtained from TGA experiments were analyzed using a computer program for non-linear regression analysis (Systat).

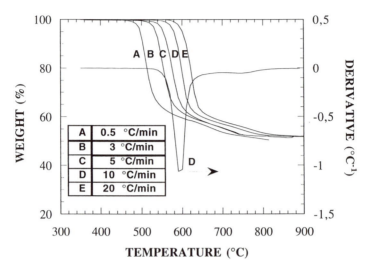

Figure 1: Weight loss of PEEK in N_2 and its derivative as a function of temperature obtained in dynamic TGA tests at different heating rates.

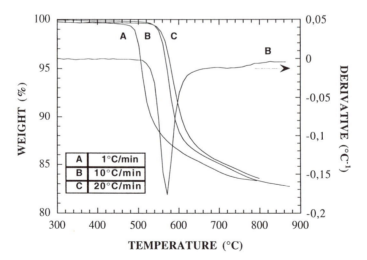

Figure 2: Weight loss of the PEEK matrix composite in N_2 and its derivative as a function of temperature obtained in dynamic TGA tests at different heating rates.

Results and Discussion

PEEK Based Materials. Results of several thermogravimetric tests performed on the PEEK matrix at different constant heating rates, in nitrogen environment, are reported in Figure 1, in terms of percentage of residual weight as a function of time. The polymer is apparently stable below 450°C. At higher temperatures the sample weight begins to diminish in a degradation process which ends above 700°C where a stable degraded material is formed. The effect of the heating rate on the degradation kinetics is manifested in the shifting of Figure 1 curves: higher heating rate curves are shifted to higher degradation temperatures. This effect is mainly dependent on the activation energy of the degradation process.

Similar degradation tests performed on the PEEK matrix composite are reported in Figure 2. In this case the effect of the carbon fibers is mainly manifested in the different residual weight of both samples. While in the first case this value (ca. 50%) is given by the remaining carbon structure of the polymer, the composite residue (ca. 83%) is also represented by the fibers that do not degrade in the nitrogen environment. A rough calculation gives, with the assumption that no carbon is lost during degradation, a fiber weight fraction of 66% which is close to the standard content (68% wt.). A second observation regards the difference between the form of the derivative curves between both materials. These curves, which can be considered as a representation of an overall degradation rate, show practically the same peak temperature. However, the degradation peak for the neat polymer is sharper than in the case of the composite. These two observations suggest a diffusion effect of the presence of the fibers which are not involved in the degradation process itself, but act as an inert filler only modifying the heat transfer in the material.

As the service life of the materials studied involves normally the presence of oxygen, weight loss tests were also performed in normal air environment. The results of a dynamic TGA test performed at 10°C/min. are reported in Figure 3 where an evident difference with tests performed in nitrogen is observed. While only the matrix is degraded in nitrogen in a carbonization process, a complete degradation of the material is achieved in air. This difference is also manifested in the difference of the derivative curve obtained in air with respect to Figures 1 and 2 results. A single peak characterizes the degradation reaction under nitrogen while a double peak is observed when the process is conducted in the presence of oxygen. In the second case the first peak is located in the same temperature range and weight loss as the one observed in nitrogen, and can be attributed to matrix degradation, while the second peak can be easily attributed to the oxidation of the carbon fibers and the carbon structure of the degraded matrix. A similar weight loss and temperature degradation interval characterizes the degradation of the matrix in both cases suggesting that the first degradation process of the matrix is given mainly by thermally controlled dehydrogenation, without strong influence of oxidation processes on the reaction yield.

While dynamic test results shown in Figures 1-3 give useful information on the temperature interval and total weight loss associated with the degradation reactions; isothermal test results are needed for kinetic analysis in order to decouple the contribution of temperature. These results are shown in Figures 4-6 for different materials and environmental conditions. The test temperatures were chosen following the dynamic results in the observed degradation interval. Higher values of the temperature produced very fast process with poor quality thermograms, and lower temperatures produced very slow processes with very long time experiments (>24 hours).

By simple inspection of weight loss results the observations from dynamic tests are confirmed. A wider degradation interval is shown by the composite (Figure 5) compared with the neat polymer (Figure 4) and the degradation process in air (Figure 6) is faster than in nitrogen (Figure 5). It must be noted that while the presence of

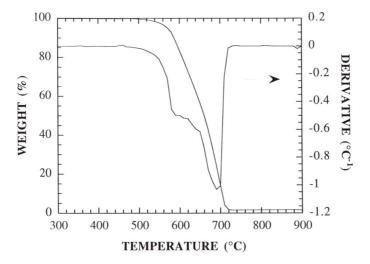

Figure 3: Weight loss of the PEEK matrix composite in air and its derivative as a function of temperature obtained in a dynamic TGA test at 10 °C/min.

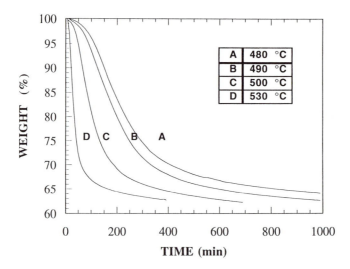

Figure 4. Weight loss of PEEK in N_2 as a function of time during isothermal TGA tests at different temperatures.

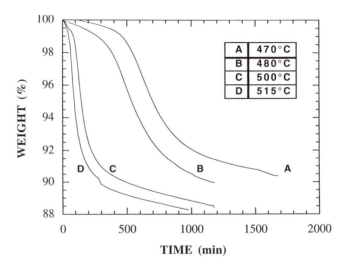

Figure 5. Weight loss of the PEEK matrix composite in N_2 as a function of time during isothermal TGA tests at different temperatures.

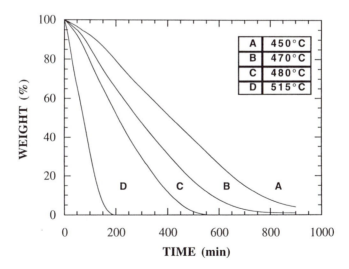

Figure 6. Weight loss of the PEEK matrix composite in air as a function of time during isothermal TGA tests at different temperatures.

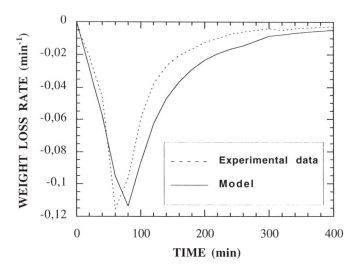

Figure 7. Model predictions and experimental results of the weight loss rate of the PEEK matrix composite during an isothermal TGA test in N_2 at 515°C.

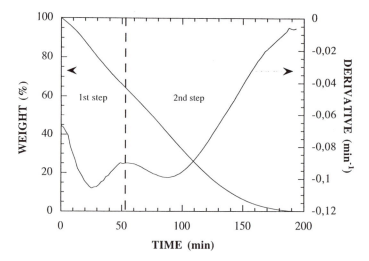

Figure 8. Reaction steps of the degradation of the PEEK matrix composite during a TGA isothermal test in air at 515 °C.

oxygen produces, as in dynamic tests, a complete disappearance of the sample, the tests performed in nitrogen produced a residue that in all cases is higher than those obtained in dynamic tests. Indeed in these cases the degradation process probably could not be completed as a consequence of the long times involved in the test. However, the form of the thermograms suggests that a degraded form is produced with a stable morphology that depends on the particular process temperature.

Also in isothermal test the difference between tests performed in nitrogen and air is manifested in the form of the weight loss rate peaks, obtained by derivation of the weight curve and reported in Figures 7 and 8. Again, a double peak is shown in the thermogram obtained in air (Figure 8) confirming the presence of two different degradation processes, while only one peak characterizes the thermogram obtained in nitrogen (Figure 7).

PEI Based Materials. Similar dynamic and isothermal TGA tests were performed on PEI under both nitrogen and air environments. In particular the results obtained in dynamic tests conducted in nitrogen are reported in Figure 9. Although the form of the weight loss curves and of their derivatives are similar to those obtained for PEEK, a significant shift of ca. 50°C to lower temperatures is observed in the derivative curve peak of PEI with respect to PEEK, suggesting a lower thermal stability of PEI in spite of its higher Tg value. This lower thermal stability is still more evident in the isothermal tests (Figure 10) which show a very fast initial degradation with the formation of a smooth plateau at long times that is a function of the test temperature. Similar results have been obtained on the PEI matrix composite. It should be reminded that PEEK can crystallize with a crystalline content of ca. 35% while PEI is amorphous. Although the degradation study has been performed well above the melting point of PEEK, the crystalline morphology also responsible its higher solvent resistance with respect to PEI, will probably increase the differences between the thermal stability of both polymers at lower temperatures below the range used in this study.

Modeling of the Degradation Kinetics

The good thermal stability of PEEK and of its carbon fiber composite is reflected in the long time scale needed to complete the degradation in nitrogen atmosphere. A typical experiment performed near 500°C lasts for about 24 hours, leading to a practical impossibility to perform experiments under controlled conditions at lower temperatures, closer to maximum service conditions (ca. 250°C). Therefore, other procedures should be followed in order to predict long time behavior of these materials.

The first approach analyzed in this research, is the development of a kinetic model that could allow the prediction of the loss weight as a function of exposure time and temperature. In order to correlate TGA experimental data with a kinetic expression, a generic degree of degradation can be defined as:

$$\alpha = (M_0-M)/(M_0-M_f), \qquad (1)$$

where M, M_0 and M_f are the actual, initial and final sample weights respectively. A typical model for a kinetic process can be expressed (4) in the following general form:

$$d\alpha/dt = K(T) f(\alpha), \qquad (2)$$

where the temperature dependence of the kinetic constant can be expressed through an Arrhenius expression:

$$K = K_0 \exp(-E/RT), \qquad (3)$$

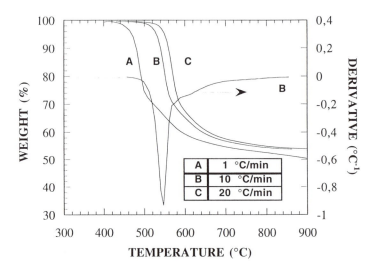

Figure 9: Weight loss of PEI in N_2 and its derivative as a function of temperature obtained in dynamic TGA tests at different heating rates.

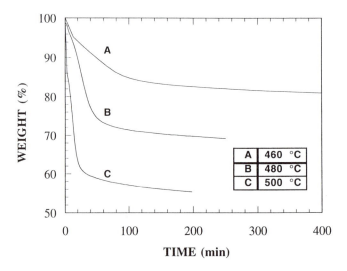

Figure 10. Weight loss of PEI in N_2 as a function of time during isothermal TGA tests at different temperatures.

where E is the activation energy. The form of the function $f(\alpha)$ depends on the form of the isothermal TGA thermogram.

Several kinetic models could fit the form of the derivative curves obtained in the experimental characterization (*15*) including a polynomial expression. However, by analogy with the results obtained in other kinetic processes like thermosets polymerization in isothermal and dynamic conditions (*16*), the so-called autocatalytic model is proposed here to analyze the degradation process of PEEK and PEI:

$$d\alpha/dt = K(T) \, \alpha^m \, (1 - \alpha)^n \qquad (4)$$

where m and n are reaction orders. In agreement with the experimental results reported in Figures 1, 2 and 9, Equation 4 predicts zero reaction rate at both process ends ($\alpha = 0$ and $\alpha = 1$).

Several methods are reported in literature to calculate the kinetic parameters of Equations 3 and 4 (*17-19*). In this work a multiple non-linear regression analysis of isothermal and dynamic results was performed using a statistical software package (Systat). The values of the model parameters obtained for the different materials studied are reported in Table I. Although the model is purely empirical and its parameters cannot be related to the real degradation mechanism, a general similarity between the behavior of the materials analyzed is observed.

Table I. Parameters of the degradation kinetic model

	$\ln(K_o)$ $[\ln(\text{min}^{-1})]$	E [KJ/mol]	m	n
PEEK/N_2	22.41	160.3	0.64	1.64
PEI/N_2	14.85	115.3	0.80	1.40
APC2/N_2	34.53	241.1	0.80	1.80

The validity of the model has been verified through comparisons between predictions and experimental results obtained in different testing conditions. Figures 7 and 11 show a good agreement for two isothermal degradation tests performed on the PEEK matrix composite and on PEI respectively. While these comparisons are expressed in terms of the degradation rate taken as directly proportional to the derivative of the isothermal TGA thermogram, Figure 12 results are expressed in terms of the degree of degradation of the PEEK matrix composite obtained as a function of time for different temperatures. A general agreement between theoretical and experimental results can be observed.

The ability of the model to represent non isothermal degradation processes has also been verified. Figure 13 shows the comparison of model and experimental results for two dynamic tests performed at different heating rates on the PEEK matrix composite. Although a general agreement has been obtained, specially for the onset and for the first part of the degradation process, some discrepancies are observed in the last part. These differences can be attributed to complex degradation mechanisms that are oversimplified when represented by a simple model like Equation 4. However, for durability studies at long term and at lower temperatures the model developed can be very useful to predict the initial part of the degradation process.

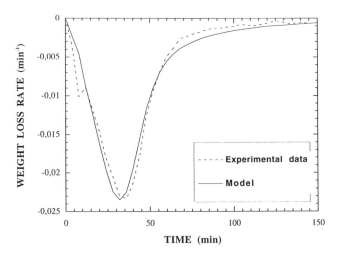

Figure 11. Model predictions and experimental results of the weight loss rate of PEI in N_2 during an isothermal TGA test in N_2 at 480°C.

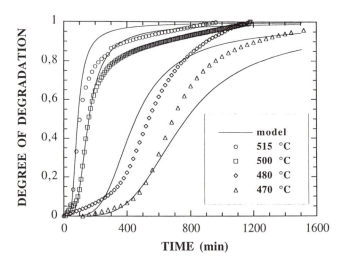

Figure 12. Model predictions and experimental results of the degree of degradation of the PEEK matrix composite in N_2 during isothermal tests at different temperatures.

Time-Temperature Correlation. A faster engineering approach to study the long term degradability of these materials can be developed considering that in practice a detectable degraded state can be represented by a weight loss of 1 or 2 % as a limit of the good functioning of the material. In the hypothesis of small weight loss it can be proved that, following Equation 4 model, the time to reach a fixed weight loss at constant temperature is given by:

$$t = \alpha^{(1-m)}/(1-m) \exp(E/RT) \qquad (5)$$

Then, plotting ln(t) vs. 1/T the experimental values should lay on a straight line as shown in Figure 14 for the PEEK matrix composite, for both levels of weight loss (1% and 2%) and for degradation in air and nitrogen . It can be noticed that the experimental data for the two different weight losses lay on parallel lines. Indeed, the lines obtained in Figure 14 by simple linear regression of experimental data are coincident with the linear behavior obtained by application of Equation 5 correlation. Therefore, in the hypothesis of small weight loss, it is possible to extrapolate from short term data the long term behavior at lower temperatures as shown in Figure 15. Indeed, the extrapolation is meaningful only in the amorphous region above the melting point also indicated in Figure 15. Below this point the fraction of crystalline material formed would certainly stabilize the polymer further with an expected deviation of the extrapolated lines to higher times. Then, the behavior represented in Figure 15 can be interpreted as a conservative limit of the estimated degradation at lower temperatures. It has to be pointed out that this kind of analysis is an engineering approximation; therefore it will be more reliable when closer is the prediction temperature to the interval used in the experimental tests.

The behavior of PEI has been also analyzed with this approach and the results are shown in Figure 16 for 1% weight loss compared with PEEK. While the values obtained for the PEEK matrix composite, both in air and in nitrogen, confirmed its good thermal stability, the shifting of PEI curves to lower temperatures is a clear confirm of its lower thermal stability. However, in both cases, the time necessary to reach a weight loss of 1 % at temperatures of the magnitude of their glass transition temperature is practically infinite. Moreover, an evident difference on the effect of the environment composition on the behavior of both materials can be observed. While the oxygen presence accelerates only slightly the degradation process of PEI, it produces a strong reduction in the degradation times of PEEK, in the temperature interval analyzed. It should also be noted that the slope of the straight lines corresponding to the same environmental conditions, are similar for both materials, indicating that their respective degradation processes are represented by similar activation energies.

Conclusions

The thermal stability and the degradation kinetics of PEEK, PEI and their continuous carbon fiber reinforced composites has been studied. The experimental results, obtained by thermogravimetric analysis performed in air and pure nitrogen, have been used to determine their degradation behavior and to develop a phenomenological kinetic model which is able to predict their degradation rate under different environmental conditions. Although the complex degradation mechanism of these materials could not be completely reproduced, the results of the model can be used to estimate the onset of the degradation process of these materials at longer times and lower temperatures than those used in the experimental characterization. The experimental characterization has shown a good thermal stability of the materials studied at their service temperatures with a better performance of PEEK with respect to PEI. Carbon fibers appear to modify slightly the degradation kinetics of the composites as a consequence of their high thermal conductivity.

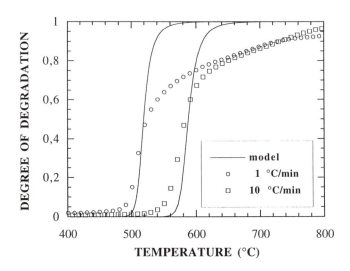

Figure 13. Model predictions and experimental results of the degree of degradation of the PEEK matrix composite in N_2 during dynamic TGA tests at two different heating rates.

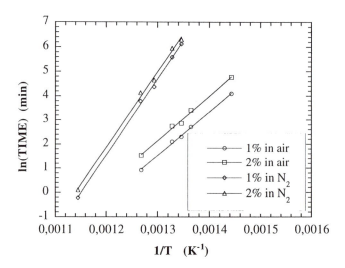

Figure 14. Time required to achieve 1% and 2 % degradation of the PEEK matrix composite in air and N_2 as a function of temperature.

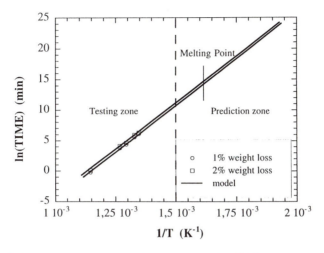

Figure 15. Prediction of long term degradation of the PEEK matrix composite at service temperatures.

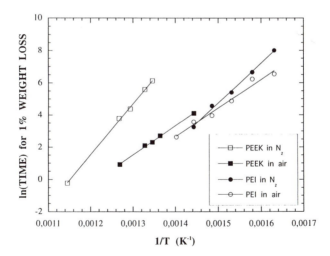

Figure 16. Time required to achieve 1% degradation of the PEEK matrix composite and PEI in air and N_2 as a function of temperature.

Literature Cited

1. Deslandes, Y.; Day, M.; Sabir, N.F.; Suprunchuk T. *Polym. Composites* 1989, 10, 360.
2. Nguyen, H.X; Ishida, H. *Polym. Composites* 1987, *8*, 57.
3. Hou, T.H.; Reddy, R.M. *SAMPE Quart.* 1991, *22(2)*, 38.
4. Skinnier, H.A. *Polym. Eng. Sci.* 1992, *32*, 17.
5. Montaudo, G.; Puglisi, C.In *Development in Polymer Degradation* ; Grassie, N., Ed.; Elsevier: London, 1987; Vol. 7.
6. Prime, R.B.; Seferis, J.C. *J. Polym Sci., Polym. Letters* 1986, *24*, 641.
7. Nam, J.; Seferis, J.C. *J. Polym. Sci, Polym Phys.* 1992, *30*, 455.
8. *Thermal Characterization of Polymeric Materials;* Turi, E.A., Ed.; Academic Press, New York, 1981.
9. Hawkins, W.L. *Polymer Degradation and Stabilization;* Springer-Verlag, Hamburg, 1983.
10. Nam, J.; Seferis, J.C. *J. Polym. Sci , Polym Phys.;* 1991, *29*, 601.
11. Levenspiel, O. *Chemical Reaction Engineering;* J. Wiley, New York, 1972.
12. Carberry; J.J. *Chemical and Catalytic Reaction Engineering;* Mc Graw Hill, New York, 1976.
13. Cogswell, F.N. *Proceedings of the 10th SAMPE European Chapter Conference;* Birmingham, U.K., 1989.
14. Kenny, J.M.; Torre, L.; Nicolais, L. *Proceedings of the 9th International Conference on Composite Materials*, Madrid, Spain, 1993.
15. Harrison, J. In *Comprehensive Chemical Kinetics;* Banford, C.; Tipper, C., Eds.; Elsevier, Amsterdam, 1980, Vol. 2, p. 377.
16. Kenny, J.M.; Trivisano, A. *Polym. Eng. Sci.* 1991, *31*, 19.
17. Flynn, J.H.; Wall, L.A. *Polym. Lett.* 1966, *4*, 323.
18. Kissinger, H.E. *Analyt. Chem.* 1957, *29*, 1702.
19. Horowitz, H.H.; Metzger, G. *Analyt. Chem.* 1963, *35*, 1464.

RECEIVED March 14, 1995

Chapter 10

Monte Carlo Simulation of Polymer High-Temperature Degradation and Fracture

N. Rapoport[1] and A. A. Efros[2]

[1]Department of Materials Science and Engineering and [2]Department of Computer Science, University of Utah, Salt Lake City, UT 84112

A Monte Carlo simulation was developed to evaluate thermo-oxidation kinetics and durability of solid polymers upon heterogeneous initiation of active radicals induced by sporadically distributed catalytic impurities. Spatial propagation of oxidation proceeded by way of the random walk of small active radicals from initiation sites. The dependence of the durability on the number of initiation centers was rather smooth for uninhibited samples but very dramatic for inhibited ones. The difference in the durability of uninhibited and inhibited samples quickly diminished with increasing number of initiation centers. Results imply that inhibitor efficiency drops drastically with increasing number of initiation centers in a polymer sample; the introduction of inhibitors, even at high concentrations, cannot suppress oxidative degradation of polymers if they contain relatively high concentrations of catalytic impurities.

The lifetimes of many polymer products are limited by oxidative degradation. Oxidative degradation is the most widespread cause of polymer failure at high temperatures in oxidative environments (e.g. air) if the mechanical stress imposed on the sample is not too high. A proper evaluation of oxidative degradation is essential both for predicting polymer lifetime and for developing new stabilizing systems.

Oxidation of polymers comprising hydrocarbon fragments is a radical chain reaction with a degenerate branching of kinetic chains. Polymer oxidation in a liquid phase (in the solution and in the melt) is reasonably well described by the regularities of a liquid-phase oxidation of low molecular weight hydrocarbons. However, oxidation of solid polymers is diffusionally controlled and cannot be described by the kinetic regularities of the liquid-phase oxidation. The limiting step of the reaction in a solid phase is either reagents' collisions or the re-orientation of reagents in a pair required for the formation of a transition complex.

Due to the diffusional control, individual steps of the kinetic scheme of the solid polymer chain oxidation do not represent elementary reactions. Rate constants of these steps are effective parameters which include corresponding diffusion coefficients. The kinetic situation becomes even more complicated if the initiation of active radicals proceeds heterogeneously, being catalyzed by impurities (e.g. polymerization catalyst residues), which is often the case in polyolefin oxidation (see

0097–6156/95/0603–0155$12.00/0

below). New techniques need to be developed to describe this complex kinetics. As long as the reaction is diffusionally controlled, Monte Carlo simulation appears to be a reasonable approach.

The Monte Carlo technique is a powerful tool for solving percolation theory problems, and is widely used for this purpose. In its essence, any mathematical method that uses the generator of random numbers uses a Monte Carlo methodology. There exists a wide variety of Monte Carlo programs customized to solving particular problems of polymer physics. However, to the best of our knowledge, it has never been used to evaluate solid polymer oxidative degradation and fracture.

Justifying the use of the Monte Carlo technique in this application is: i) the solid polymer oxidation rate being controlled by the radical microdiffusion from the initiation centers, and ii) polymer fracture resulting from the formation of a percolation cluster of microdefects. Here we report the results of the first attempt to simulate solid polymer oxidative degradation and failure using a Monte Carlo technique.

Oxidation chemistry. In polyolefins and polymers with hydrocarbon fragments, oxidative degradation proceeds as a radical reaction with branched kinetic chains. The oxidation scheme is given below. It comprises four main stages:

Reaction	Rate controlling factor	
Initiation:		
$RH + O_2 \longrightarrow R\cdot + HO_2\cdot$	W_0	(0)
Propagation:		
$R\cdot + O_2 \longrightarrow RO_2\cdot$	k_1	(1)
$RO_2\cdot + RH \longrightarrow ROOH + R\cdot$	k_2	(2)
Branching		
$ROOH \longrightarrow dR\cdot$	k_3	(3)
Termination		
$R\cdot + R\cdot \longrightarrow$ inactive products	k_4	(4)
$R\cdot + RO_2\cdot \longrightarrow$ inactive products	k_5	(5)
$RO_2\cdot + RO_2\cdot \longrightarrow$ inactive products	k_6	(6)

Here, W_0 is the rate of primary initiation of active radicals, and the k_i's are the effective rate constants of the corresponding stages. Depending on the oxygen pressure, the termination proceeds either by disproportionation of two peroxide radicals (reaction 6) or with the participation of alkyl radicals (reactions 4 and 5). In the presence of inhibitors, the main termination reaction involves peroxide radical decay in interaction with inhibitor molecules or inhibitor radicals:

$RO_2\cdot + InH \longrightarrow ROOH + In\cdot$	k_7	(7)
$RO_2\cdot + In\cdot \longrightarrow$ inactive products	k_8	(8)

In polyolefins, radical initiation (reaction 0) is catalyzed by impurities, which are presumably polymerization catalyst residues (*1-7*). The importance of impurities was confirmed in a study that applied X-ray photoelectron spectroscopy to the investigation of highly oxidized sites in slightly oxidized polypropylene (N.R. and I. Sapozhnikova, unpublished). Sites of deeper oxidation were revealed by the incubation of oxidized films in an SO_2 atmosphere. Also, fracture sites of the samples oxidized under tensile stress were compared with sites away from fracture zones. Highly oxidized sites and fracture sites contained trace amounts of transition metals and high concentrations of silicon, which is used in polymerization catalyst support systems. None of these were found in unoxidized sites or away from the fracture sites. It was concluded that the initiation of polypropylene oxidation was catalyzed by the polymerization catalyst residues. Thus, oxidation initiation occurred at pre-existing, randomly distributed centers.

Kinetic chain propagation and termination reactions in solid polymers are controlled by diffusion (7). Macroradicals in a solid phase have a very restricted motion. In this case, spatial propagation of the oxidative degradation from initiation centers is believed to proceed by way of translational diffusion of low molecular weight radicals formed in the decomposition of peroxide or alkoxyl macroradicals (7,). This process involves three stages:

- *The formation of a small radical, upon decomposition of a peroxide radical:*

$$RO_2\cdot \longrightarrow r\cdot \qquad\qquad k_d \qquad\qquad (9)$$

- *Translational diffusion of r·*   *diffusion coefficient D*   (10)

- *Radical exchange reaction:*

$$r\cdot + RH\ (O_2) \longrightarrow rH + R\cdot\ (RO_2\cdot) \qquad\qquad k_p' \qquad\qquad (11)$$

The last reaction results in the transformation of low molecular weight radicals to macroradicals. This reaction is one of the modes of kinetic chain propagation. Due to this reaction sequence, a peroxide macroradical, $RO_2\cdot$, moves from the initial localization site by a distance r which depends on the diffusion coefficient of a small radical, D, and on the exchange reaction rate constant, k_p':

$$r = (Dt)^{1/2} = (D/k_p'[RH])^{1/2}$$

The rate of the spatial propagation of the oxidation can be expressed as:

$$W(sp) = k_d(D/k_p'[RH])^{1/2}$$

where k_d is the frequency of these events.

Polymer chain scission results from hydroperoxide thermal decomposition (see below). Therefore, the higher the hydroperoxide concentration and decomposition rate, the higher the rate of polymer degradation. At some degree of local oxidative degradation, microcracks are generated; as oxidation proceeds, microcracks grow, coalesce and eventually give rise to a catastrophic crack.

Special Features of the Oxidation of Solid Polymers. Thermo-oxidation of solid polymers is controlled by the micro-diffusion of radicals, which proceeds almost exclusively in the amorphous phase of semi-crystalline polymers and is sensitive to polymer morphology and mechanical stress (7). In the initial stage of oxidation, reaction is confined to the initiation centers, and the volume of polymer actually involved in oxidation is orders of magnitude lower than the total sample volume. At this initial stage, two processes proceed simultaneously: the degree of oxidation of the regions already involved in oxidation increases, and the reaction front moves by radical diffusion.

Due to the heterogeneous radical initiation and the diffusional control of kinetic chain propagation and termination, oxidation kinetics of solid polymers does not follow the regularities of a liquid-phase oxidation, and new techniques are required to model the behavior. Here, a Monte Carlo simulation is used to evaluate the kinetics of this process. Basic kinetics parameters used in the calculation refer to polypropylene (PP) oxidation at 130° C. Diffusion of low molecular weight substances and polypropylene thermo-oxidation proceeds exclusively in the amorphous phase; the oxidation temperature of polypropylene is well above its T_g. Therefore, the first-order rate constant, $k_p[RH]$, of reaction 11 was assumed to be close to that in a liquid phase (namely, on the order of 10^2 /s), resulting in the lifetime of a small "jogger" radical before immobilization, t_r, being on the order of 10^{-2} s. One tenth of a "jogger" radical lifetime, $t = 10^{-3}$ s, was chosen as a time unit, t.u., for the simulation.

The diffusion coefficient of low molecular weight radicals, r·, was assumed to be close to that of gases in PP, of about 10^{-5} cm^2/s at 130°C. During a "jogger" radical's lifetime it moved an average distance $(Dt_r)^{1/2}$, which is about 3 μm. The width of one amorphous layer in PP is about 300 Å (8). Thus during a "jogger" radical's lifetime, it crossed at least 100 amorphous layers which corresponded to 10 lattice sites.

Oxidation resulted in the formation and accumulation of microdefects (microcracks); merging of microcracks into a percolation cluster was taken as the moment of failure. At this moment, the simulation stopped and the corresponding time was considered the polymer durability.

Algorithm of the Simulation.

Radical Generation. A specified number of initiation centers, N, with random coordinates was inserted in a 2-D square lattice. Durability experiments were done on the 50 x 50 lattice; kinetic experiments were done on the 25 x 25 lattice.

With each specific number of initiating centers, experiments were repeated 10 times using differing sets of initiation center coordinates. Mean-square values of durability, t, and standard deviations, s, were evaluated, as well as kinetics of radical and microcrack accumulation and residual concentration of an inhibitor.

Movable, low molecular weight "jogger" radicals, r·, were generated at initiation centers at a rate of $W_o = 1$/t.u.. The number of initiation centers was varied from 5 to 100. The chosen initiation rate corresponded to a relatively high *average* radical initiation rate on the order of 10^{-6} to 10^{-5} mol/kg.s. Experimental data on radical generation rates in PP autooxidation vary widely, ranging from 10^{-8} to 10^{-5} mol/kg.s at 130°C; the highest initiation rates were attributed to the presence of catalytic impurities (9).

Radical Diffusion. "Jogger" radicals diffused through the lattice sites by way of a random walk. After a specified number of jumps, a (a = 10 in this simulation), radical r· became immobilized; the immobilization simulated the transformation of a low molecular weight "jogger" radical r· into the non-movable macroradical R· (RO$_2$·) by reaction 11.

Polymer Degradation. Polymer degradation proceeds with the participation of immobilized macroradicals. The key event in polymer degradation is the decomposition of hydroperoxide, ROOH, formed in reaction 2. Thermal decomposition of hydroperoxide is accompanied by polymer degradation which proceeds upon the decomposition of alkoxyl radical RO· according to one of the following reactions:

$$ROOH \longrightarrow [RO· + OH·] \longrightarrow \delta R· + {>}C{=}O + \text{macromolecule scission} \qquad (12)$$

$$ROOH + RH \longrightarrow [RO·+ R·] + H_2O \longrightarrow \delta R· + {>}C{=}O + \text{scission} \qquad (13)$$

$$\text{ROOH} + \text{ROOH} \longrightarrow [\text{RO}_2\cdot + \text{RO}\cdot] + \text{H}_2\text{O} \longrightarrow \delta\text{R}\cdot + >\text{C=O} + \text{scission} \qquad (14)$$

New radicals, R·, are generated during hydroperoxide decomposition. Hence, reactions 12 - 14 are degenerate kinetic chain branching reactions.

For simplicity in what follows, all types of macroradical (alkyl R·, alkoxyl RO· and peroxide $\text{RO}_2\cdot$) are designated as R·. Only immobilized macroradicals participate in polymer degradation. The role of small "jogger" radicals, r·, is to provide for the spatial propagation of the degradation reaction by way of a random walk through the polymer matrix followed by intermittent immobilization according to reaction 11.

To simulate these events, the "jogger" radical was transformed into a macroradical after 10 diffusion steps. The macroradical stayed immobilized for the time t_R, during which time the polymer degradation proceeded in a corresponding site. After t_R time units, the macroradical decomposed, re-generating a "jogger" radical; the point of macroradical immobilization was treated as an origin for the new "jogger" radical's trajectory. After radical departure from its immobilization site (or after macroradical decay in collision with a "jogger" radical, see the following section), another radical could become immobilized at the same site, thus increasing the degree of its degradation. When the total time spent by macroradicals at a given lattice site attained t_d time units, this site was considered a microcrack.

Bimolecular Radical Decay (Uninhibited Oxidation). Radicals "killed" each other at collisions. Bimolecular radical decay proceeded in a collision of two "jogger" radicals jumping simultaneously to the same lattice site or in a collision of a "jogger" radical with a macroradical at the site of its localization.

Monomolecular Radical Decay (Inhibited Oxidation). Each lattice site was "charged" with a specific number, h, of inhibitor molecules, which "killed" radicals according to reactions 7 and 8. Therefore, 2h radicals could decay in collisions with inhibitor molecules at each lattice site. One inhibitor molecule was consumed for every two radicals killed. Spatial propagation of the degradation reaction from site to site could proceed only after the inhibitor was completely consumed at a particular lattice site. In our calculation, the inhibitor concentration, h, varied from 0 (which corresponded to uninhibited oxidation) to 50.

Polymer Failure. Polymer degradation in amorphous layers for the time t_d resulted in microcrack formation. Microcracks accumulated and eventually merged into a percolation cluster, causing a catastrophic failure. Positions of the defects were collected in intervals of 1 time unit; at each simulation step, the formation of clusters of defects was tracked, starting from the left edge of the lattice and moving to the right. Only vertical and horizontal contacts of defects were considered. The length of the simulation run was determined by the time required to form the percolation cluster of defects, and was taken to be the sample durability.

Degenerate Chain Branching. After macroradical R· spent t_R time units at some lattice site, this site was considered a potential source of new radicals due to the accumulation of hydroperoxide (reaction 2) followed by the hydroperoxide decomposition (reactions 12 - 14); the probability, β, of one-time generation of two consecutive radicals at these sites was included into the simulation.

Results.

Dependence of the predicted durability on lattice size was checked by running 25 x 25, 50 x 50, and 100 x 100 lattices with the same concentration of initiation centers. The durability was found to be independent of lattice size. Random variations in predicted durabilities were checked by running 10-experiment sets under each

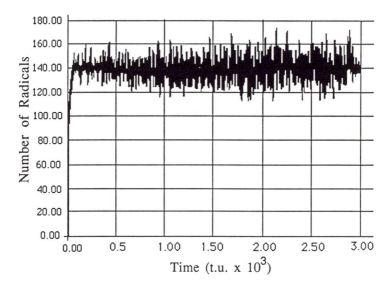

Figure 1. Kinetic curve of radical accumulation in uninhibited oxidation; 50 initiation centers in a 25 x 25 lattice.

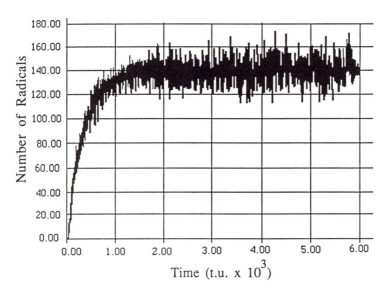

Figure 2. Kinetic curve of radical accumulation in inhibited oxidation; 50 initiation centers in a 25 x 25 lattice; 20 inhibitor molecules per a lattice site.

condition investigated. At higher concentrations of initiation centers, the scatter in durability was very small (less than 2%). At lower concentrations, the variation from case to case was larger.

Kinetics. Typical kinetic curves of radical accumulation in the absence and presence of an inhibitor are presented in Figures 1 and 2 for the same number of initiation centers (50 in a 25 x 25 lattice). The number of inhibitor molecules, h, is zero in Figure 1 and 50 in Figure 2. As follows from comparing Figures 1 and 2, stationary radical concentration is the same in uninhibited and inhibited samples; however the initial rate of radical accumulation is much higher in uninhibited samples; it decreases proportionally to the inhibitor concentration. It is interesting that in inhibited samples the stationary radical concentration was established long before all the inhibitor was consumed, which means that at the sites where the inhibitor was consumed, the degradation proceeded at the same rate as in uninhibited samples.

As could be expected, the rate of radical accumulation and stationary radical concentration depended on the number of initiation centers. The dependence of stationary radical concentration on the number of initiation centers (and hence on the *average* initiation rate) differed from one characteristic of the liquid-phase oxidation: semi-logarithmic rather than square root dependence was observed (Figure 3).

These kinetic results appear quite reasonable which suggests that due to the micro-diffusional control of the reaction, the kinetics of solid-phase polymer oxidation can be successfully simulated using the Monte Carlo technique. Typical kinetic curves of microcracks accumulation are presented in Figures 4 and 5 respectively for uninhibited and inhibited samples, for n = 50. The curves start from 50 on the Y axis because sites of initiation are considered as defects. Accumulation of defects proceeded with an induction period followed by a period of autoacceleration, but slowed down before the sample failure (Figure 5). The introduction of an inhibitor considerably prolonged the duration of the induction period but affected to a much lesser extent the rate of defect accumulation during the autoacceleration stage (compare Figures 4 and 5).

Durability. Durability of the samples depended on the number of initiation centers (Figure 6) and on the concentration of an inhibitor (Figure 7). Figure 6 shows that the dependence of the durability on the number of initiation centers is rather smooth in uninhibited samples but very dramatic in inhibited ones. The difference in the durability of uninhibited and inhibited samples quickly diminished with increasing number of initiation centers. It implies that inhibitor efficiency drops dramatically with increasing number of catalytic impurities in a polymer sample.

As follows from the data of Figure 7, no matter how much of an inhibitor is introduced into a polymer sample, it is hardly possible to increase its durability if a substantial concentration of catalytic impurities is present (n = 100, lower curve). Decreasing the number of initiation centers renders an inhibitor much more effective: at n = 20 (Figure 7, upper curve) the introduction of an inhibitor resulted in about a two order of magnitude increase in polymer durability. Simulation results imply that the elimination of initiation centers offers a promising way of increasing inhibitor efficiency and prolonging polymer lifetime.

Residual Concentration of an Inhibitor. Not all the inhibitor was consumed at the moment of sample failure. Residual concentration of an inhibitor in a failed sample did not depend on the initial concentration of an inhibitor, but dropped sharply with increasing number of initiation centers in a sample (Figure 8). These results could have implications for polymer recycling.

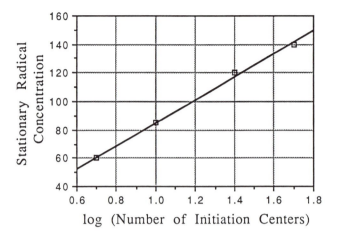

Figure 3. Stationary radical number in a 25 x 25 lattice vs. the number of initiation centers.

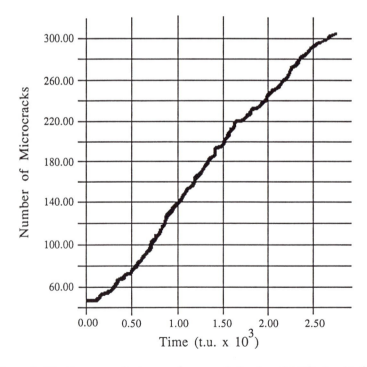

Figure 4. Kinetic curve of microcrack accumulation in uninhibited oxidation; 50 initiation centers in a 25 x 25 lattice.

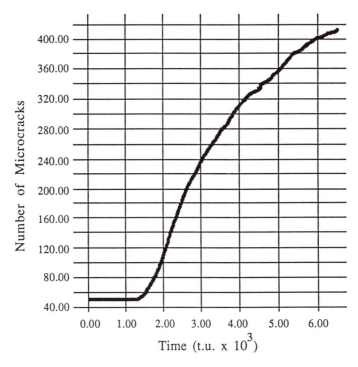

Figure 5. Kinetic curve of microcrack accumulation in inhibited oxidation; 50 initiation centers in a 25 x 25 lattice; 20 inhibitor molecules per a lattice site.

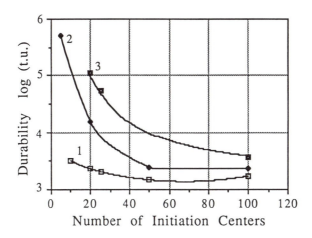

Figure 6. Durability vs.the number of initiation centers; the number of inhibitor molecules per a lattice site in a 50 x 50 lattice: 0 (1), 5 (2), 20 (3).

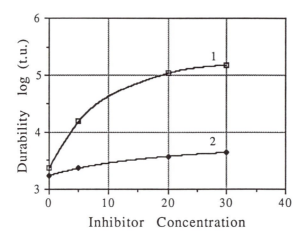

Figure 7. Durability vs. inhibitor concentration; the number of initiation centers in a 50 x 50 lattice: 20 (1) and 100 (2).

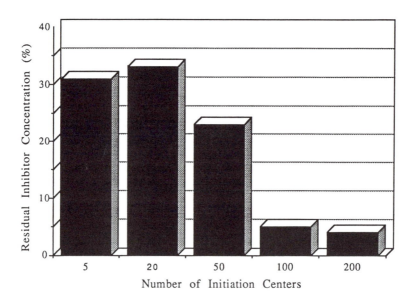

Figure 8. Residual inhibitor concentration in a failed sample vs. the number of initiation centers in a 50 x 50 lattice.

Literature Cited.

1. Richters, P. *Macromolecules,* **1970,** *3,* 262.
2. Knight, L.B.; Calvert, P.D.; and Billingham, N.C. *Polymer,* **1985,** *26,* 1713.
3. Billingham, N.C. *Macromol. Chem., Macromol. Symp.1989,* *28,* 145.
4. Celina, M.; George, G.A.; and Billingham, N.C. *Polym. Prepr., Am. Chem. Soc., Div. Polym. Chem.,* **1993,** *34 (2),* 262.
5. Rapoport, N. *Intern. J. Polymeric Mater.,* **1993,** *19,* 93.
6. Rapoport, N. et al. *Intern. J. Polymeric Mater.,* **1993,** *19,* 101.
7. Popov, A. A.; Rapoport, N.Ya, and Zaikov, G.E. *Oxidation of Stressed Polymers* ; Gordon and Breach Science Publ.: New York, 1991.
8. Gesalov, M.A.; Kuksenko, B.S.; and Slutsker, A.I. *Physika Tverdogo Tela (Russian),* **1971,** *13,* 445.
9. Denisov, E.T. *Russian Chemical Reviews,* **1978,** *47,* 1090.

RECEIVED May 15, 1995

APPLICATIONS AND NEW MATERIALS

Chapter 11

Novel High-Performance Thermosetting Polymer Matrix Material Systems

G. W. Meyer, S. J. Pak, Y. J. Lee, and J. E. McGrath

Department of Chemistry and National Science Foundation Science and Technology Center, High Performance Polymeric Adhesives and Composites, Virginia Polytechnic Institute and State University, Blacksburg, VA 24061-0344

The synthesis and characterization of phenylethynyl terminated triaryl phosphine oxide containing polyimide oligomers, poly(arylene ether sulfone triaryl phosphine oxide) copolymers and highly processable thermosetting poly(ether imide) systems is described. The polyimide oligomers were controlled molecular weight systems endcapped with 3-phenylethynyl aniline or novel 4-phenylethynylphthalic anhydride. The poly(arylene ether) copolymers were synthesized in high molecular weight with controlled amounts of a novel co-monomer which contained pendant aryl amino groups which were subsequently functionalized with 4-phenylethynylphthalic anhydride. The phenylethynyl groups were employed to afford a higher curing temperature (380-420°C) which widens the processing window between Tg and the cure temperature as compared to the well studied acetylene endcapped systems. Thermally cured samples flow under low pressure during processing, display excellent solvent resistance, and exhibit very high char yields as determined by TGA, e.g., the polyimides displayed about 50% char at 700°C in air.

Advances in thermosetting resins for functional and structural applications which offer properties similar to those of thermoplastics but with improved solvent resistance and better high-temperature dimensional stability are needed (1). Thermoplastics and thermosets have both been widely used for composite applications in high performance materials. Thermoplastics have the advantages of minimal chemistry during processing, as well as ductility, and may be recycled after the fabrication process. However, amorphous thermoplastics typically have a major deficiency in solvent resistance. Thermosets on the other hand do offer excellent solvent resistance and a large body of research is available on thermosetting systems. However, these resins may be too brittle (1) and are difficult to alter or recycle once crosslinking has taken place. As a result, by combining the advantageous properties of both thermoplastics and thermosets, one attempts to produce a material with processability, adequate durability and solvent resistance.

0097-6156/95/0603-0168$12.00/0
© 1995 American Chemical Society

Thermoplastic polyimides such as Ultem® 1000 can be designed to exhibit excellent thermal stability and mechanical properties. A number of polyimide systems have been developed that are well suited for use as matrix resins, adhesives, and coatings for high performance applications in the aerospace and electronics industries (2-4). However, as mentioned above, one major disadvantage of thermoplastic systems is their poor solvent resistance. There is a growing demand for polyimide resins that are processable, yet solvent resistant, and that will withstand very high temperatures, perhaps in excess of 700°F (371°C) for extended periods of time. There is an additional need for moderately high temperature adhesives and composites matrices for the proposed high speed civil transport airplane. This has provided a strong incentive for developing processable polyimide systems that could be commercialized to meet these demands. As a result, we have focused on the incorporation of 3-phenylethynyl aniline (5), <u>1</u>, or 4-phenylethynylphthalic anhydride (6), <u>2</u>, as the endcapping agents, which cure at ~360 - 420°C.

Phenylethynyl Aniline (1)   4-Phenylphthalic Anhydride (2)

Hergenrother earlier reported that phenylethynyl groups randomly spaced along a quinoxaline backbone produces a thermally induced crosslinking reaction occuring at approximately 430-450°C (7) It was previously reported that the use of 4-phenylethynylphthalic anhydride-functionalized oligomers cure upon heating to afford a highly crosslinked network that is highly solvent resistant and that can show exceptional thermooxidative stability (6). With the use of these higher curing endcaps, polyimides may be utilized that exhibit Tg's in excess of 400°C after curing and still allow sufficient processing time at temperatures well above the oligomer Tg before the system is fully cured. By designing highly processable backbone structures such as poly(ether imides), these reactive oligomers possess relatively low melt viscosities above their Tg and thermally cure without the evolution of volatile byproducts.

Polyarylene ether sulfones such as UDEL® are excellent engineering thermoplastics which are widely used in a variety of applications since they offer excellent mechanical properties, high Tg's and are hydrolytically and thermooxidatively stable. However, these materials are also solvent sensitive, especially under stress and undergo creep under load at elevated temperatures approaching Tg. Jensen et al. (8) have synthesized phenylethynyl-containing monomers based on the 3F systems (9) which can be polymerized to afford pendant acetylenic groups along the polymer backbone. In this paper, we describe a different approach where we have synthesized poly(arylene ether) copolymers which contain pendant aryl amine moieties. These pendant amines were then functionalized to afford maleimide (10) or imide phenylethynyl groups to give solvent resistant highly ductile poly(arylene ether) resins.

In addition to crosslinking the polyimide and poly(arylene ether) polymers to improve solvent resistance, we have also incorporated phosphorus into the backbone of these polymeric materials. Triarylphosphine oxide-containing polymers have also been shown in our laboratories to be excellent adhesives to metal substrates (11). The good adhesive characteristics of these systems may be a function of the polar interaction of the phosphine oxide moiety with the surface of the metal substrate. In addition, phosphine oxide-containing polyimides and poly(arylene ethers) show excellent thermal stability; surpassing the thermal stability of more conventional aromatic-based systems such as polycarbonates or polyesters (12,13). Even more desirable, phosphine oxide based polymers are excellent flame retardant materials due to their self-extinguishing nature (14). Therefore, we have devoted considerable efforts towards developing highly processable, high Tg thermosetting phosphine oxide containing polymers to satisfy these demands.

Experimental

Materials. N-methyl pyrrolidinone (NMP) and N,N'-dimethyl acetamide (DMAc) were obtained from Fisher and were vacuum distilled over P_2O_5 and stored under N_2 prior to use. 1,2-Dichlorobenzene (o-DCB) was obtained from Fisher, and used as received. Toluene was obtained from Fisher and vacuum distilled over calcium hydride and stored under N_2 prior to use. Bis[4-(3,4-dicarboxyphenoxy)phenyl]propane dianhydride or Bisphenol A dianhydride (BPA-DA) was kindly donated by General Electric. BPA-DA was monomer grade material; however, it was dried at ~175°C under vacuum prior to use. Pyromellitic dianhydride (PMDA) was kindly donated by Allco, and was not further purified. Hexafluoroisopropylidene-2,2-bis (phthalic acid anhydride) (6FDA) was kindly donated by Hoechst Celanese and was not further purified. 4,4'-Oxydianiline (ODA) was obtained from Chriskev and was vacuum sublimed twice prior to use. m-Phenylene diamine (m-PDA) was obtained from Aldrich and was vacuum sublimed twice prior to use. m-Bis(aminophenyl)phenylphosphine oxide (m-BAPPO) was synthesized as previously reported (15). 3-Phenylethynyl aniline (16c) and 4-phenylethynylphthalic anhydride (4-PEPA) (6b) were prepared as previously reported. Monomer grade Bisphenol-A was provided by Dow Chemical and was used as received. Bis (4-fluorophenyl) phenylphosphine oxide (DFTPPO) was synthesized according to a literature procedure (17) and was purified by vacuum distillation. Amino DFTPPO was synthesized as previously reported (10).

Phosphine Oxide Phenylethynyl Aniline Terminated Polyimide Synthesis.

Polymers were synthesized via the previously described ester-acid route (16). A representative polymerization conducted for the synthesis of a PMDA/m-BAPPO 9000 g/mole oligomer is provided as follows: Monomer grade PMDA, 74.96 grams (3.4366×10^{-1} moles) was charged to a 2000 mL 3-neck round bottom flask equipped with an overhead stirrer, nitrogen inlet, thermometer, reverse Dean-Stark trap and reflux condenser. The reactor was heated with a silicone oil bath. Absolute ethanol (7-10 mL per gram of dianhydride) and ~12 mL triethylamine were then introduced. The mixture, was then refluxed with stirring until a clear solution was obtained. When the distillation of ethanol ceased, the trap was filled with o-dichlorobenzene. Next, 9.7943 grams (5.0682×10^{-2} moles) of 3-phenylethynyl aniline was then charged to control molecular weight as a solution in 100 mL of NMP. The diamine, m-BAPPO, 156.74 grams (3.1825×10^{-1} moles), was then dissolved in 200 mL NMP and charged into the reaction vessel. An additional 280 mL of NMP was then added to the reaction vessel. An azeotroping solvent, o-dichlorobenzene, 145 mL, was then added to provide a solids content of 25% wt.:vol.. The reaction mixture was then slowly heated to

170-185°C and held for ~15 hours, after which time the viscous polymer solution was cooled to room temperature and coagulated by slowly dripping the polyimide solution into an excess of methanol in a high speed blender. The polymer was collected by vacuum filtration, washed with methanol and then anhydrous diethyl ether, air dried 6-8 hours and vacuum dried at ~170°C for 24 hours.

Conversion of a Poly(Arylene Ether) Copolymer Containing 10% Pendant Amines (10) to Phenylethynyl Phenylimides. To a four neck round bottom flask equipped with a mechanical stirrer, nitrogen inlet, thermometer, and a reverse Dean Stark trap with a condenser was added 5.5000 g (1.2×10^{-3} moles) of the pendant amino poly(arylene ether phosphine oxide) copolymer (10) along with 55 ml (20 wt% solids) of NMP. The contents were allowed to stir at room temperature until a homogeneous solution was obtained. Then, 0.5770 g (2.0×10^{-3} moles) of 4-phenylethynylphthalic anhydride was charged into the flask and allowed to stir at room temperature for an additional 4 hours, after which 11 ml of o-dichlorobenzene was added and the solution was heated to 180°C for 8 hours. The resulting polymer with pendant phenylethynyl phenylimides was cooled and then coagulated into methanol two times and dried in a vacuum oven at 100°C for 8 hours.

Phenylethynylphthalic Anhydride Terminated Polyimide Synthesis. Polymerizations were conducted as follows for the synthesis of BPA-DA/m-PDA oligomers with a number average molecular weight $\bar{M}n$ of 7,000 g/mole provided as an example: 6.817×10^{-3} moles (3.548 grams) monomer grade BPA-DA and 1.1545×10^{-3} moles (0.287 grams) of 4-phenylethynylphthalic anhydride to control molecular weight and endgroups were charged to a 3-neck round bottom flask equipped with a magnetic stirrer, nitrogen inlet, thermometer, reverse Dean-Stark trap and reflux condenser which was heated in an oil bath. Absolute ethanol, 7-10 ml per gram of dianhydride and ~3 mL triethylamine were then introduced and the mixture was refluxed with stirring until a clear solution was obtained. When the distillation of ethanol ceased, the trap was filled with o-DCB. The diamine, m-PDA, 7.3945×10^{-3} moles (0.800 grams) was then added into the reaction vessel, followed by NMP (11 ml) and o-DCB (3 ml) (80:20 vol.:vol.) to provide a solids content of 20 % wt.:vol.. The reaction mixture was then heated to 170-185°C for 20 hours, after which time the polymer solution was coagulated by slowly dripping the cooled polyimide solution into methanol in a high speed blender. The polymer was collected by vacuum filtration, washed with methanol and then with anhydrous diethyl ether. It was then air dried 6-8 hours and vacuum dried at ~170°C for 24 hours.

Cured phenylethynyl terminated films were prepared by compression molding at 380°C for 90 minutes at 40-400 psi depending on the molecular weight unless otherwise specified. Tensile property samples were cut with a dog bone die conforming to ASTM D368 type V. Tensile shear specimens were prepared by solution coating an adhesive tape (112 glass) containing 2 percent volatiles which was cured for 1 hour at 350°C at 50 psi in between titanium tensile shear speciments (6A1-4V). All values are the average of 4-5 runs per specimen.

Characterization. FTIR spectra obtained with a Nicolet MX-1 FTIR spectrometer on powder samples were used to determine the presence of imide groups in the polymers. Proton NMR spectra obtained with a Varian Unity 400 spectrometer were also used to confirm the polymer structure. Intrinsic viscosity measurements of the imide samples were performed in NMP or chloroform at 25°C using a Canon-Ubbelohde viscometer. Gel Permeation Chromatography

(GPC) measurements were performed on a Waters 150-C ALC/GPC with a viscosity detector; $\overline{M}n$ and $\overline{M}n/\overline{M}w$ values for the polyimide samples were determined using universal calibration (18). Potentiometric titrations were performed using a MCI GT-05 automatic titrator to determine the number average molecular weight per amine group. The amine functionalities were titrated with HBr (0.025N). Glass transition temperatures and cure exotherms were determined by differential scanning calorimetry (DSC) using a Perkin Elmer DSC 7 Differential Scanning Calorimeter. Scans were run at a heating rate of 10°C/minute; reported values were obtained from a second heat after quick cooling. Thermogravimetric analyses (TGA) were performed on a Perkin Elmer TGA 7 at 10°C/minute in air for dynamic scans. Tensile property measurements were taken on an Instron Model 1123 following the ASTM D638 method. The measurements were taken at a strain rate of 0.05 in/min at room temperature. Parallel plate scans were performed on a Bohlin Rheometer in nitrogen at a heating rate of 3°C/minute using a frequency of 1 Hz.

Results and Discussion

Phosphine Oxide-Containing Phenylethynyl Aniline Terminated Polyimides.
The phenylethynyl terminated phosphine oxide containing polyimide matrix resins have been synthesized using the ester-acid route in high yields by a process previously described (16). The polyimides were derived from the PMDA/m-BAPPO system where the resulting polyimides have been controlled to the targeted number average molecular weight, $\overline{M}n$, of ~7000 and 10,000 g/mole. The intrinsic viscosities in NMP at 25°C were 0.17 and 0.20 dl/g, respectively. The $\overline{M}n$ as determined by GPC, was 6700 g/mole for the 7000 g/mole sample. Scheme 1 illustrates the general synthetic scheme for the phenylethynyl aniline terminated polyimides.

The Tg values of the polyimides were determined by DSC before and after network formation, as shown in Table I. The Tg of the 7,000 g/mole polymer before curing on the first scan was 240°C. Curing the polyimide for 90 minutes at 380°C in flowing air, increased the Tg to 264°C. The thermal stability was also investigated by thermal gravimetric analysis (TGA) in air by comparing the polymer before cure, after 90 minutes at 350°C and after 90 minutes at 380°C. As seen in Figure 1, the initial weight loss is essentially unchanged by the thermal treatment. However, the char yield, (which is often correlated with flame resistance) determined at 700°C for each of the three curing schedules was influenced. The char yield after curing for 90 minutes at 380°C, increased substantially to 49 weight percent. These initial results suggest that the polymer network exhibited improved char formation when cured for 90 minutes at 380°C.

Table I. Thermal Analysis of Triarylphosphine Oxide Containing Phenylethynyl Aniline End Capped Polyimides

System	Target MW x 10³ (g/mole)	5% Wt. Loss* (°C)	Tg₁ (°C)	Tg₂ (°C)
PMDA/m-BAPPO	7.0	529	236	264
	10.0	517	239	255

*	=	Heating rate of 10°C/min in air, TGA of the in situ cured oligomer
Tg₁	=	Tg before cure, heating rate of 10°C/min in nitrogen
Tg₂	=	Tg after cure, heating rate of 10°C/min in nitrogen

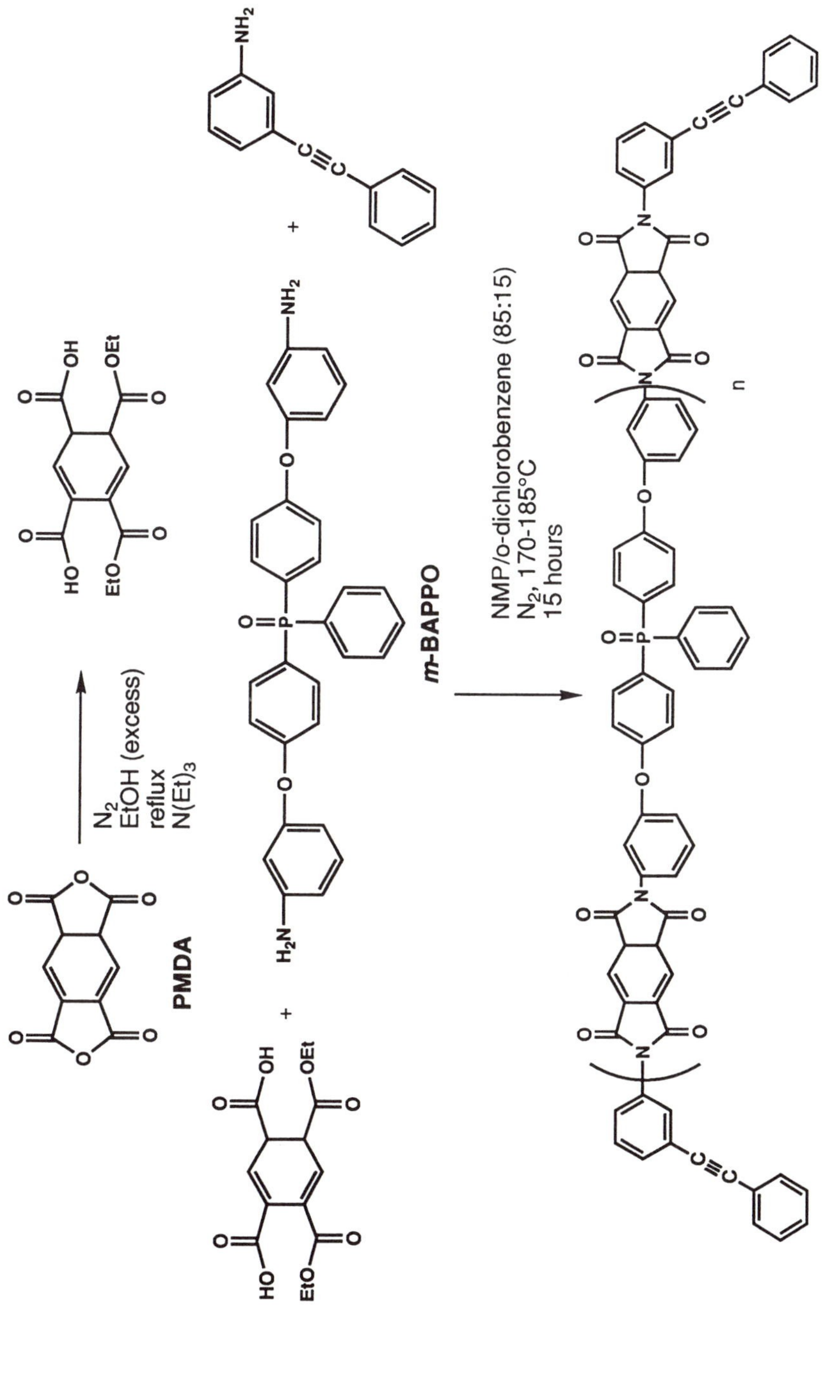

Scheme 1. Synthesis of Phenylethynyl Aniline Terminated Phosphine Oxide Polyimides

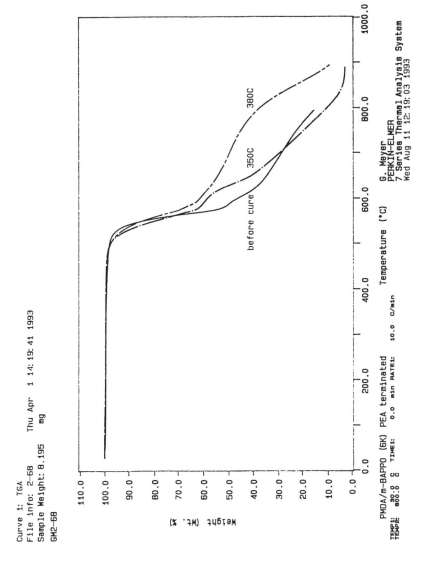

Figure 1. TGA Thermograms of PMDA/*m*-BAPPO Phenylethynyl Aniline Terminated Polyimides Prepared Under Different Cure Schedules.

Preliminary studies to examine the gel fraction of the cured samples have also been performed. Films cured at various times and temperatures were extracted with chloroform (which is a solvent for the linear material) for 5 days. They were then dried for 5 days at 100°C under vacuum. As Table II exhibits, the sample cured at 380°C for 90 minutes had the highest gel fraction at 98%. By contrast, samples cured for 90 minutes at 350°C exhibited a high, but significantly lower gel fraction of 89%.

Table II. Influence of Time and Temperature on the Gel Fraction of ($\overline{M}n$) = 7,000 g/mole PMDA/*m*-BAPPO Phenylethynyl Cured Polyimide Oligomer Networks

Temperature (°C)	Time (minutes)	Gel Content %
350	60	67
350	90	89
380	30	90
380	60	94
380	90	98

Solvent Extraction Experiments - Each sample was extracted with chloroform for 5 days, and then dried 5 days at 100°C under vacuum.

Mechanical testing has also been initiated to measure the tensile properties of the cured polyimide. Some data based on samples cured at 350°C for 90 minutes are in the general range expected for polyimides, and are shown in Table III.

Table III. Tensile Behavior of ($\overline{M}n$) = 7,000 g/mole PMDA/*m*-BAPPO Phenylethynyl Cured Polyimide Networks

Modulus (ksi) (MPa)	403 (2779)
Yield Stress	11.1 (76.5)
Yield Strain (%)	4.0

Samples were cured at 350°C for 90 minutes.

Furthermore, initial evaluation of the composite materials based on this polyimide system have been provided by NASA Langley and these results are shown in Table IV.

The flow properties and cure window, defined as the interval between Tg and T_{gel} have been examined by rheological parallel plate studies. The viscosity-temperature behavior and the real and imaginary shear moduli have been plotted against temperature as shown in Figures 2 and 3. The gel point may be established to be ~380°C as judged from the crossover point of the real and imaginary (shear and loss moduli), G' and G".

Phosphine Oxide-Containing Poly(arylene ether)s. Phenylethynylphthalic anhydride was also successfully reacted with pendant aryl amines on copolymers of poly(arylene ether phosphine oxides) to produce pendant phenylethynyl phenylimides. The latter were subsequently thermally cured to afford networks with an increase of Tg up to 17°C, as shown in Table V. Scheme 2 illustrates the general synthetic scheme for the poly(arylene ether) phosphine oxide copolymers as was reported earlier (10) and Scheme 3 illustrates the general synthetic scheme

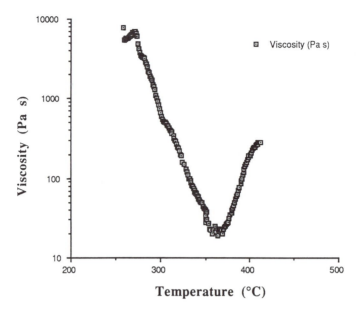

Figure 2. Viscosity vs. Temperature Profile of PMDA/*m*-BAPPO
Phenylethynyl Aniline Terminated Polyimide (7,000 g/mole).

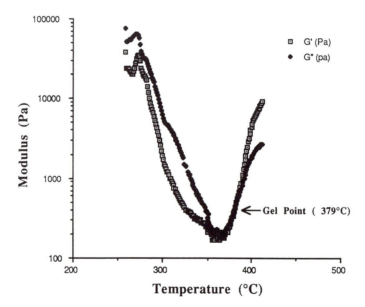

Figure 3. Influence of Temperature on the Loss and Storage Moduli of
PMDA/*m*-BAPPO Phenylethynyl Aniline Terminated Polyimide
(7,000 g/mole).

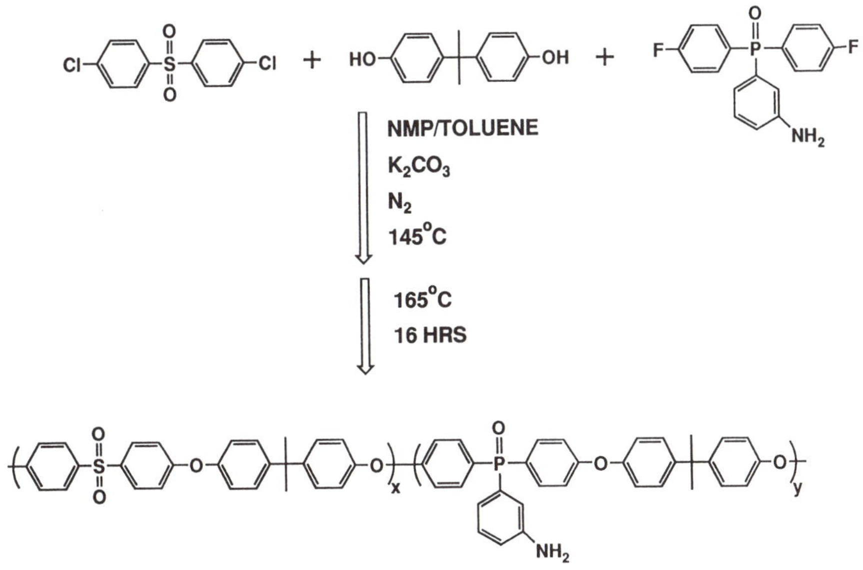

Scheme 2. Synthesis of Poly(arylene ether) Phosphine Oxide Copolymer with Pendant Amines.

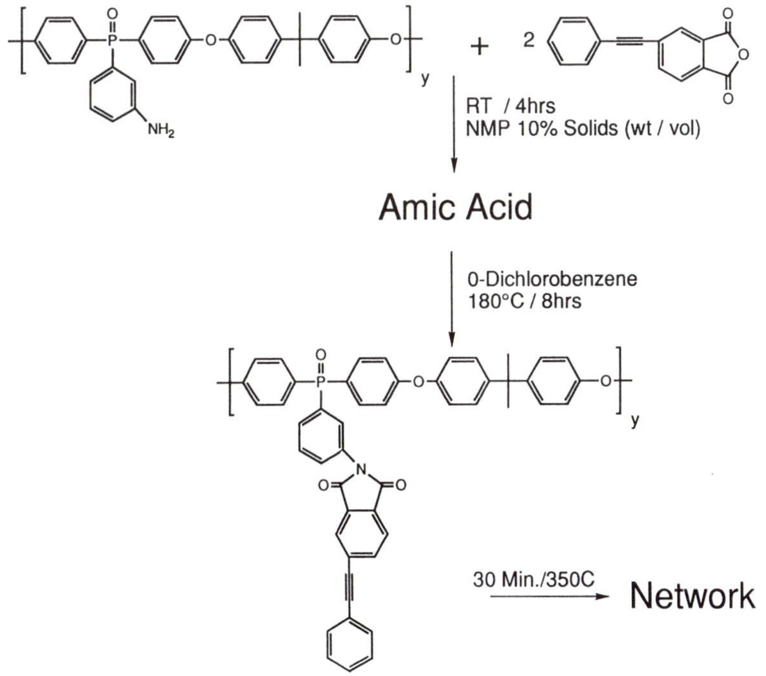

Scheme 3. Conversion of Poly(arylene ether) Phosphine Oxide Copolymer to Pendant Phenylethynyl Phenylimides.

for the conversion of poly(arylene ether) phosphine oxides to pendant phenylethynyl phenylimides.

Table IV. Mechanical Testing of PMDA/*m*-BAPPO Phenylethynyl Aniline Terminated Polyimide 7,000 g/mole, ($\overline{M}$n) oligomer[a]

<u>Ti/Ti Tensile Shear Strength</u> (cure conditions of 1 hour at 350°C, 50 psi)

Test Temperature	Strength (psi) (MPa)
RT	4760 (32.8)
177°C	3927 (27.1)

<u>IM-7 Unidirectional Composite Flexural Properties</u>
(cured 0.5 hr at 300°C--150 psi, 1 hr at 375°C, 100 psi)

Test Temperature	Flexural Strength (ksi) (MPa)	Modulus (Msi) (MPa)
RT	219 (1510)	21.4 (147,553)
177°C	143 (986)	18.2 (125,489)

Test Temperature	Compression Strength (psi)
RT	182

[a]Data courtesy of NASA Langley/William and Mary

Table V.   Characterization of Poly(Arylene Ether) Containing Pendant Phenylethynylimide Moieties

Polymer	[η] 25°C CHCl₃ (dl/g)	<$\overline{M}$n> Amine (g/mole)	TG(°C) Uncured	Tg(°C) Cured	% Gel Fraction*
PEPO Control	1.09	---	---	---	0
5% Phenylethynyl	1.01	9,700	201	201	93
10% Phenylethynyl	0.99	4,700	201	207	98
20% Phenylethynyl	0.81	2,500	201	218	99

*Soxhlet Extraction in Chloroform for three days.

The results from the gelation studies show that the phenylethynyl groups react nicely to afford networks with increased solvent resistance.

Phenylethynyl Phthalic Anhydride Terminated Polyimide Resins. 4-Phenylethynylphthalic anhydride terminated polyimide matrix resins have been synthesized using the ester-acid route (16) in high yields. The materials were based on 6FDA and ODA. The oligomers were prepared at three different molecular weights ranging from 3000 to 15,000 g/mole. Scheme 4 illustrates the

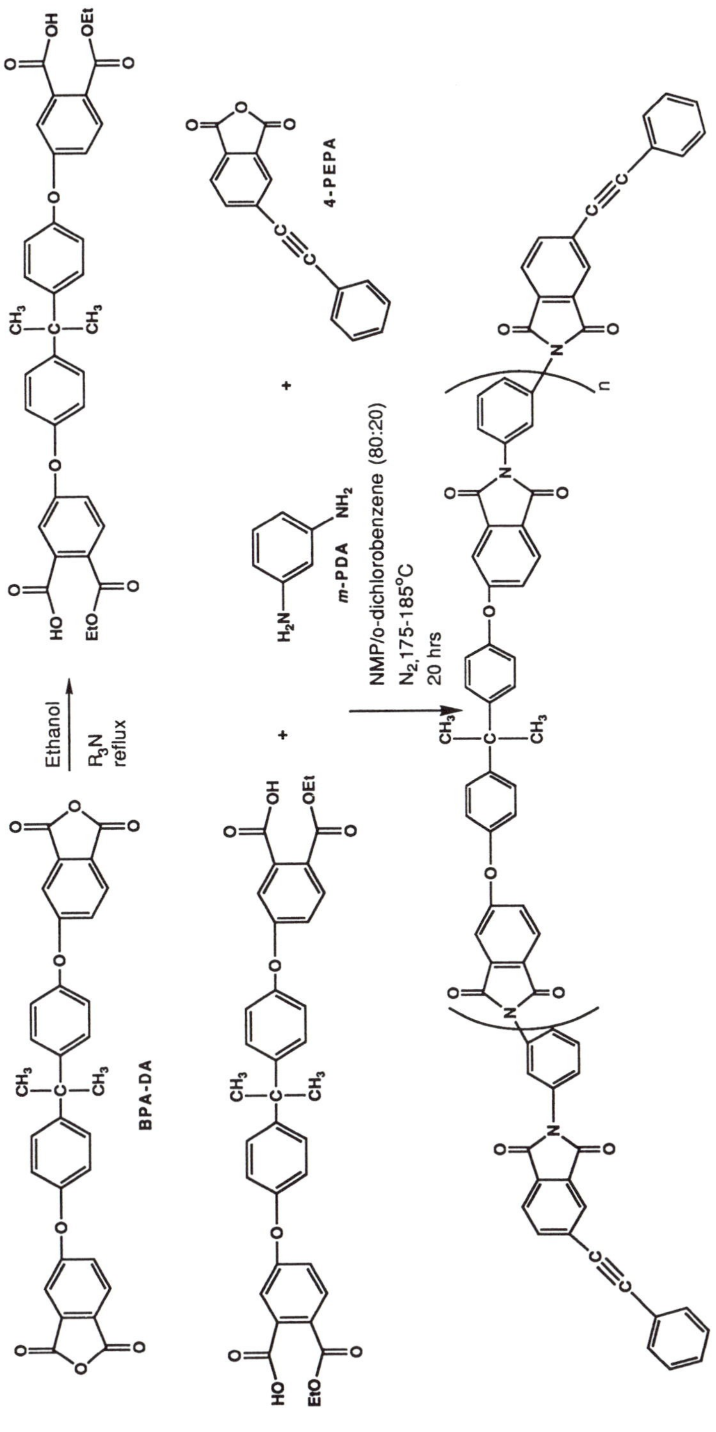

Scheme 4. Synthesis of Phenylethynylphthalic Anydride Terminated Polyimides.

general synthetic scheme using phenylethynylphthalic anhydride to prepare phenylethynylphthalimide terminated polyimides.

Molecular weight control was achieved by offsetting the monomer stoichiometry in favor of the diamine and utilizing 4-phenylethynylphthalic anhydride as a endcapping agent, as shown in Table VI. The intrinsic viscosity values listed correspond very well with the anticipated molecular weight of the 6FDA/ODA polyimide systems. These materials appeared to be completely imidized, as judged by the strong infrared imide absorptions at 1780 cm^{-1}, 1730 cm^{-1}, 1370 cm^{-1}, and 710 cm^{-1} and absence of absorptions attributable commonly to poly(amic-acids). All of the polyimide oligomers (even the 3000 g/mole system) formed tough, creasable films after curing (380°C, 90 minutes, 40-400 psi).

Table VI. Molecular Weight Control Data for 4-Phenylethynylphthalic Anhydride Terminated Polyimides

System	Calculated Number Average Average Molecular Weight ($\overline{M}n$ x 10^3)	$[\eta]^{NMP}$dL/g 25°C
6FDA/ODA	3.0	0.15
	10	0.32
	15	0.41

Table VII lists the glass transition temperatures before curing (255-288°C) and after curing (289-310°C) were consistent with the Tg (~300°C) of the high molecular weight linear systems. The onset of exothermic crosslinking occured at ~380-420°C, as judged by DSC (10°C/minute) measurements.

Table VII. Thermal Analysis of 4-Phenylethynylphthalic Anhydride Functionalized Polyimides[*]

System	Target ($\overline{M}n$) x 10^3	Tg_1 (°C)	Tg_2 (°C)	TGA_1 (°C)	TGA_2 (°C)
6FDA/ODA	3.0	255	310	555	587
	10	275	308	551	577
	15	288	289	546	----

*	=	Heating rate of 10°C/min in nitrogen, TGA of oligomer
Tg_1	=	Tg before cure, heating rate of 10°C/min in nitrogen
Tg_2	=	Tg after cure, heating rate of 10°C/min in nitrogen
TGA_1	=	5 % Wt. Loss before cure, measured in air at 10°C/minute
TGA_2	=	5 % Wt. Loss after cure, measured in air at 10°C/minute

The thermal stability of these phenylethynyl terminated polyimides was also investigated by subjecting the cured samples to dynamic thermal gravimetric analysis (TGA) and 5% weight loss data in air show values of 577-587°C, (Table VII). Thermally cured samples also display good solvent resistance when immersed in DMAc at ambient temperature for 1 week, as judged by stress-crack and short term swelling resistance.

Phenylethynylphthalic Anydride Terminated Poly(Ether Imides). Scheme 4 illustrates the general synthetic scheme using phenylethynylphthalic anhydride to prepare phenylethynylphthalimide-terminated poly(ether imides) polyimides. *Meta*-phenylene diamine was the only diamine employed in the phenylethynylphthalic anhydride terminated polyimide systems, so that a direct comparison could be made with the linear system (Ultem®) in terms of increased glass transition temperatures and solvent resistance after curing of phenylethynyl endgroups.   Molecular   weight   control   was   achieved   utilizing   4-phenylethynylphthalic anhydride as an endcapping agent, as shown in Table VIII with molecular weights ranging from 3,000 to 30,000 g/mole. The intrinsic viscosity values listed correspond very well with the anticipated molecular weight of the BPA-DA/*m*-PDA systems.

Table VIII. Molecular Weight Characterization of Phenylethynylphthalic Anhydride Poly(Ether Imides)

System	Target $\overline{M}n$ x 10^3	$[\eta]$ dL/g (NMP,25°C)
BPA-DA/*m*-PDA	3.0	0.18
	7.0	0.28
	30.0	0.47

Glass transition temperatures before cure increased with increasing molecular weight with values increasing from 192 to 218°C as shown in Table IX. Films of the imide oligomers were formed by melt pressing the polyimides between steel platens (~20-40 psi pressure) and were cured for 90 minutes at 380°C. After thermal crosslinking, the glass transition temperatures increased significantly up to a maximum of 233°C for the 3,000 g/mole system and this molecular weight also displayed the largest increase in Tg after curing. The DSC curve of the 3,000 g/mole oligomer before and after thermal crosslinking is shown in Figure 4. The cured samples also exhibited excellent thermal stability in air as also shown in Table IX.

Table IX.   Thermal Characterization of Phenylethynylphthalic Anhydride Poly(Ether Imides)

System	Target $\overline{M}n$ x 10^3	Tg before cure (°C)*	Tgafter cure (°C)*	TGA 5% Wt. Loss (°C)
BPA-DA/ *m*-PDA PEPA	3.0	192	233	547
	7.0	205	227	540
	30.0	218	218	539

*   =   Heating rate of 10°C/min in nitrogen
TGA =   5 % Wt. Loss after cure, measured in air at 10°C/minute

Samples were cured at 380°C for 90 minutes at 40-400 psi pressure.

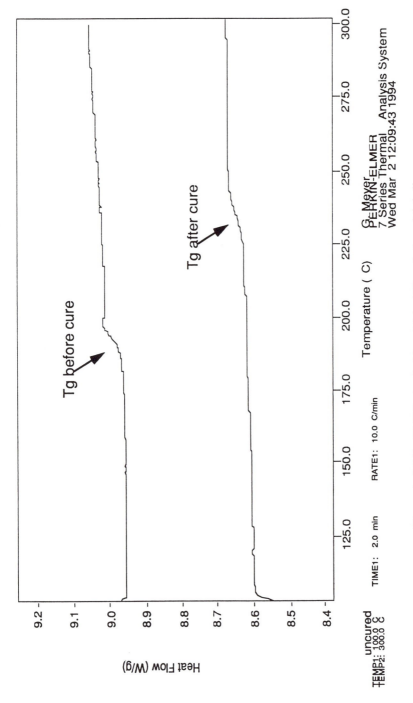

Figure 4. DSC of 3,000 g/mole *Meta*-Linked Phenylethynylphthalic Anhydride-Terminated Polyimide Oligomer Before and After Curing.

All of the polyimide oligomers (even the 3000 g/mole system) formed tough, creasable films by compression molding after curing. Soxhlet extractions were also performed on the cured samples where they were subjected to refluxing chloroform for 5 days. As shown in Table X, high (>95%) gel fractions were obtained upon thermally curing the phenylethynylphthalic anhydride-terminated poly(ether imide) systems. Even the 30,000 g/mole 4-PEPA terminated polymer was substantially crosslinked as demonstrated by a 78% gel fraction. Whereas, the thermoplastic Ultem® is completely soluble in chloroform.

Table X. Soxhlet Extraction Data in Chloroform of Phenylethynylphthalic Anhydride Poly(Ether Imides)

Polymer System	Target $\overline{M}n \times 10^3$	% Gel Fraction*
BPA-DA/*m*-PDA PEPA	3.0	96
	7.0	84
	30.0	78

* = Samples were extracted for 5 days and then dried for 4-5 days at 100-150°C.

Conclusions

Triaryl phosphine oxide polyimide oligomers endcapped with phenylethynyl aniline were synthesized via a "one pot" solution imidization involving ester-acid and diamine monomers to yield highly imidized, controlled molecular weight phenylethynyl functionalized polyimides which had a desirably wide processing window. Upon curing, insoluble, highly crosslinked films with the anticipated Tg values were obtained that exhibited excellent thermooxidative stability and high char yields. Also, phenylethynyl phenylimides pendant to the backbone of poly(arylene ether phosphine oxides) were synthesized by reaction of phenylethynylphthalic anhydride with precursor poly(arylene ether) pendant aryl amines. The resulting polymers were thermally cured to afford tough networks which displayed increased Tg values and greatly improved solvent resistance. Polyimide and poly(ether imide) oligomers endcapped with 4-phenylethynylphthalic anhydride were synthesized to yield controlled molecular weight phenylethynyl functionalized polyimides. Thermal curing at ~380°C produced insoluble, highly crosslinked, but yet creasable films with the anticipated Tg values which exhibit excellent thermooxidative stability. The successful behavior is related to the wide processing window (>150°C) afforded by the phenylethynyl endcap.

Current and Future Studies

Efforts are currently directed at the synthesis of additional 4-phenylethynyl terminated polyimide systems. In addition further studies aimed at elucidating the network formation and cure chemistry using solid state Magic-Angle [13]C Nuclear Magnetic Resonance are ongoing. Fabrication of structural adhesives for metal and polymer matrix carbon fiber composites is also in progress.

Acknowledgments

The authors would like to gratefully acknowledge the National Science Foundation and Technology Center for partial support under contract DMR-9120004 and additional support of the project by DARPA, as administered by the ARO. In addition, a fellowship from the Adhesive and Sealant Council and generous support from the Gencorp Foundation is greatly appreciated. The authors also wish to thank Dr. Steve Wilkinson and Mr. Paul Hergenrother of NASA Langley Research Center and the College of William and Mary for the adhesive and composite test data. The authors thank Dr. Jon Rich and the General Electric Co. for providing the BPA dianhydride.

Literature Cited

1. C. A. Arnold, P. M. Hergenrother and J. E. McGrath, "An Overview of Organic Polymeric Matrix Resins for Composites," Composites: Chemical and Physicochemical Aspects, T. L. Vigo and B. J. Kinzig, Eds., VCH 1992.
2. 'Polyimides: Materials, Chemistry and Characterization', C. Feger, M. Khojasteh, and J. E. McGrath, Editors, Elsevier, 1989.
3. 'Polyimides', D. Wilson, P. Hergenrother, and H. Stenzenberger, Editors, Chapman and Hall, 1990.
4. M. E. Rogers, T. M. Moy, Y. J. Kim and J. E. McGrath, *Mat. Res. Soc. Symp. Proc.*, 264, 13-29 (1992).
5. F. W. Harris, A. Pamidimukkala, R. Gupta, S. Das, T. Wu and G. Mock, *J. Macromol. Sci.*, A21 (8 and 9), 1117-1135, 1984.
6. a) J. G. Smith and P. M. Hergenrother, *Polym. Prepr.*, 35 (1), 353 (1994), *Polymer (London)* (accepted); b) G. W. Meyer, T. E. Glass, H. J. Grubbs, and J. E. McGrath, *Polym. Prepr.*, 35 (1), 549 (1994); G. W. Meyer, T. E. Glass, H. J. Grubbs, and J. E. McGrath, *J. Poly. Sci.* (accepted), 1994; c) F. W. Harris, J. A. Johnston, and T. Takekoshi, *Polymer (London)* (accepted), 35, 000, 1994.
7. P. M. Hergenrother, *Macromolecules*, 14, 898, 1981.
8. B. J. Jensen, P. M. Hergenrother, and G. Nwokogu, *J. M. S. - Pure Appl. Chem.*, A30(6 & 7), 449, 1993.
9. M. E. Rogers, M. H. Brink, A. Brennan, H. Marand, and J. E. McGrath, *Polymer (London)*, 34 (4), 849, 1993.
10. S. J. Pak, G. D. Lyle, R. Mercier and J. E. McGrath, *Polymer (London)*, 34 (4), 885, 1993.
11. T. H. Yoon, Y. J. Lee, A. Gungor, C. D. Smith, and J. E. McGrath, "Adhesion Study of Phosphorus Containing Polyimides and Polyethers," Adhesion Society Proceedings, Hilton Head, SC, Vol. 15, p. 200, Spring, 1992.
12. A. Gungor, C. D. Smith, J. Wescott, S. Srinivasan, and J. E. McGrath, *Polym. Prepr.* 32 (1), 172, 1991; J. Wescott, Ph.D. Thesis (1993).
13. C. D. Smith, H. J. Grubbs, H. F. Webster, J. P. Wightman, and J. E. McGrath, *P.M.S.E.* 65, 108, 1991; *High Performance Polymers*, 4, 211, 1991.
14. H. J. Grubbs, C. D. Smith and J. E. McGrath, *P.M.S.E.* 65, 111, 1991; J. E. McGrath, C. D. Smith, H. J. Grubbs, A. Gungor, J. Wescott, S. C. Liptak, and P. A. Wood, Proceedings of the International Conference for the Promotion of Advanced Fire Resistant Aircraft Interior Material, Atlantic City, NJ, p. 159-174, February 1993.
15. A. Gungor, C. D. Smith, J. Wescott, S. Srinivasan, and J. E. McGrath, *Polym. Prepr.* 32 (1), 172, 1991.

16. a) T. Moy, Ph.D. Thesis, 1993; b) T. Moy, C. D. DePorter and J. E. McGrath, *Polymer (London)*, **34** (**4**), 819, 1993; c) G. W. Meyer, J. Saikumar, and J. E. McGrath, *Poly. Prepr.*, **34** (2), 540, 1993.
17. S. Hirose, K. Nakamura, T. Hatakeyma and H. Hatakeyama, Sen-I Gakkaishi, **44**, 563 (1988).
18. M. Konas, M. E. Rogers, T. M. Moy, A. R. Schultz, T. C. Ward, and J. E. McGrath, *J. Poly. Sci. Physics*, in press (1993).

RECEIVED November 7, 1994

Chapter 12

Properties and Potential Applications of Poly(arylene ether benzimidazole)s

J. W. Connell[1], J. G. Smith, Jr.[2], and P. M. Hergenrother[1]

[1]Materials Division, NASA Langley Research Center,
Hampton, VA 23681−0001
[2]Department of Chemistry, Virginia Commonwealth University,
Richmond, VA 23284

As part of a NASA program on high performance polymers for potential aerospace applications, poly(arylene ether benzimidazole)s (PAEBIs) are under evaluation. The polymers are prepared by the aromatic nucleophilic displacement reaction of 5,5'-bis[2-(4-hydroxyphenyl)benzimidazole] with activated aromatic difluorides in a polar aprotic solvent using potassium carbonate at elevated temperatures under nitrogen. Based on preliminary screening of various material properties, two polymers were selected for more extensive evaluation. An isophthaloyl-containing PAEBI was evaluated for potential use as an adhesive and composite resin matrix . Controlled molecular weight versions of this material exhibited a good combination of processability and high unidirectional composite properties up to 232°C. This polymer also exhibited excellent adhesion to copper and polyimide, and moderate adhesion to titanium. A phosphine oxide-containing PAEBI was screened for potential use on spacecraft in low Earth orbit where resistance to atomic oxygen is required. Thin films of this material were exposed to an oxygen plasma under vacuum and the weight loss of the film was monitored as a function of exposure time. Relative to uncoated Kapton®HN film, the phosphine oxide-containing poly(arylene ether benzimidazole) film exhibited significantly lower weight loss rates. In addition, this material exhibited a non-linear weight loss rate as compared to Kapton® HN film which exhibited a linear weight loss rate. Upon exposure to the oxygen plasma, the phosphine oxide-containing poly(arylene ether benzimidazole) formed a phosphate-type surface coating as evidenced by x-ray photoelectron spectroscopy.

As part of a program at NASA directed towards high performance/high temperature polymeric materials for potential aerospace applications, poly(arylene ether benzimidazole)s (PAEBIs) are under evaluation. The PAEBIs were prepared by the reaction of novel bis(4-hydroxyphenylbenzimidazole)s with activated aromatic dihalides.[1-5] The molecular weights of the polymers were controlled by offsetting the stoichiometry and endcapping with 2-(4-hydroxyphenyl)benzimidazole. Based on preliminary screening, two polymers exhibited a favorable combination of properties and were subsequently selected for more extensive evaluation.

One polymer which contains an isophthaloyl group (Figure 1) was evaluated for potential use as a film, adhesive and composite resin matrix . This material displayed promise as a thin adhesive film for use in the fabrication of microelectronic circuitry where adhesion to copper and polyimide is required. Organic polymeric materials with a unique combination of properties that include initial solubility in polar aprotic solvents, thermal stability, adhesion to copper and polyimide, prevention of copper ion migration, acceptable dielectric constant and high mechanical properties are needed in the microelectronics industry. Materials of this type can be used to improve upon current microelectronic component fabrication processes and thereby reduce overall costs. In collaboration with a major U.S. electronics company, this material exhibited excellent properties in studies directed towards use in the fabrication of microelectronic circuitry.

The phosphine oxide-containing PAEBI[6] (AOR-PAEBI, Figure 1) was evaluated for oxygen plasma and atomic oxygen (AO) resistance. AO resistant materials are needed for applications on spacecraft in low Earth orbit (LEO) where AO is prevalent. AO severely degrades most organic materials and can cause catastrophic erosion and mass loss in relatively short time periods.[7-10] This issue is of concern primarily in LEO where AO is present in sufficient concentration and at sufficiently high energy levels to cause degradation. Certain perfluorinated polymers, such as fluorinated copoly(ethylene propylene), have exhibited good resistance to AO in both ground based and space flight exposure experiments. However, simultaneous exposure to AO and ultraviolet (UV) radiation dramatically increases the rate of degradation of this material.[11,12] Coatings of inorganic oxides such as aluminum oxide[13], silicon oxide[13], chromium oxide[14] and indium-tin oxide[15] as well as decaborane[16] based coatings, have been shown to protect organic materials from AO erosion. To provide maximum protection, the coatings need to be ~500-2000 Å thick, relatively uniform and defect free. Although coatings have been used successfully to protect materials on spacecraft from AO, they possess some inherent shortcomings. The application of thin, uniform coatings to complex shapes can be difficult and quality control must be performed to assure the absence of pinholes and defects. Generally, the coatings are readily abraded so care must be exercised during the fabrication of components.

Presently there is concern that on missions of long duration, impacts by micrometeoroids and debris which are prevalent in LEO will damage the protective coatings and expose the underlying material which will subsequently be eroded by AO. Polymeric materials with inherent AO resistance offer an added margin of safety. If desired, these materials could be coated for added AO protection, however if the coating is imperfect or somehow damaged either in LEO or during fabrication, the exposed material would erode at a significantly slower rate as compared to the materials currently in use. Potential applications for polymer films on spacecraft include flexible solar array substrates and multi-layer thermal insulation blanket material.

The results of these two evaluations as well as potential applications for these materials are discussed.

Experimental

Starting Materials. 2-(4-Hydroxyphenyl)benzimidazole (mp 280°C)[4], 5,5'-bis[2-(4-hydroxyphenyl)benzimidazole] (mp 389°C)[4] and 1,3-bis(4-fluorobenzoyl)benzene (mp 178-179°C)[17] were prepared as previously described. Bis(4-fluorophenyl)phenyl phosphine oxide was used as-received from Daychem Laboratories, Dayton Ohio. High purity grade N,N'-dimethylacetamide (DMAc) and N-methyl-2-pyrrolidinone (NMP) were used as-received from Fluka Chemical Corp. Anhydrous potassium carbonate (99+%) and toluene (reagent grade) were purchased from commercial sources and used as-received.

Isophthaloyl-Containing PAEBI (3 mole % Stoichiometric Offset). Into a 2L three necked round bottom flask equipped with nitrogen inlet, thermometer, mechanical stirrer, Dean Stark trap and condenser were placed 5,5'-bis[2-(4-

hydroxyphenyl)benzimidazole]   (68.79g,   0.165mol),   1,3-bis(4-fluorobenzoyl)benzene (54.63g, 0.17mol), 2-(4-hydroxyphenyl)benzimidazole (2.14g, 0.010mol), pulverized anhydrous potassium carbonate (59.8g, 0.43mol), dry DMAc (600 mL, 18.3 % solids) and toluene (250 mL). The mixture was heated to 140-150°C for 7.5 hr, toluene was removed and the temperature was subsequently increased to 155-160°C. After ~1.25 hr the viscous reaction mixture was diluted with 600 mL hot DMAc (10.0 % solids) and stirring continued at 155-160 °C. The viscous reaction mixture was diluted again with 400 mL hot DMAc (7.7 % solids) after an additional 2.75 hr. Stirring was continued for an additional 0.5 hr at 155-160°C and the reaction mixture was cooled. The viscous solution was precipitated in a water/ acetic acid (10/1) mixture, washed successively in hot water and methanol and dried at 110°C to provide a light tan polymer (115.2 g, 97 % yield) with a glass transition temperature (T_g) of 276°C and an inherent viscosity (0.5% solution in DMAc at 25°C) of 1.42 dL/g.

Phosphine Oxide-Containing PAEBI (3 mole % Stoichiometric Offset). Into a 5L three neck round bottom flask equipped with a mechanical stirrer, thermometer, nitrogen inlet, Dean Stark trap and condenser were placed 5,5'-bis[2-(4-hydroxyphenyl)benzimidazole] (295.9g, 0.707mol), bis(4-fluorophenyl)phenyl phosphine oxide (229.1g, 0.729mol), 2-(4-hydroxyphenyl)benzimidazole (9.20g, 0.0437mol), pulverized anhydrous potassium carbonate (242g, 1.75mol), toluene (500 mL) and NMP (2.2L, 18.2% solids). The reaction mixture was heated under nitrogen to ~145°C for ~12 hr to remove water from the system. The toluene was subsequently removed and the temperature was increased to ~175-185°C. After ~5 hr the viscous solution was diluted with additional hot NMP (1.4L, 12.0% solids) and the reaction was allowed to continue for an additional 2.5 hr. The viscous solution was allowed to cool and was precipitated into water/acetic acid mixture (10/1) and washed successively in hot water and methanol. The tan polymeric powder was air dried overnight and subsequently placed in a forced air oven at 130°C. The polymeric powder was obtained in near quantitative yield and exhibited a T_g of 365°C and an inherent viscosity of 0.77 dL/g.

Films. DMAc solutions (15% solids) of the polymers were centrifuged, the decantate doctored onto clean, dry plate glass and dried to a tack-free form in a low humidity chamber. The films on glass were stage-dried to ~330°C for the isophthaloyl-containing polymer and 365°C for the phosphine oxide-containing polymer in flowing air or nitrogen. Thin film tensile properties were determined according to ASTM D882 using at least four specimens per test condition.

Molded Specimens. Powdered, controlled molecular weight endcapped isophthaloyl-containing PAEBIs (approximately 10 g) were compression molded in a 3.2 cm^2 stainless steel mold by heating to 365°C under 2.4 MPa and held for ~1 hr. Miniature compact tension specimens (1.6 cm X 1.6 cm X 0.95 cm thick) were machined from the moldings and subsequently tested to determine their fracture toughness (K_{Ic}, critical stress intensity factor) according to ASTM E399 using four specimens per test condition. G_{Ic} (critical strain energy release rate) was calculated using the mathematical relationship, $G_{Ic} = (K_{Ic})^2/E$, where E is the modulus of the material.

Adhesive Specimens. Adhesive tape was prepared by multiple coats of 112 E glass with an A-1100 finish secured on a frame with DMAc solutions of controlled molecular weight endcapped PAEBI and subsequently dried to 200°C after each coat. Titanium (Ti)-to-Ti (6Al-4V) tensile shear specimens with a Pasa-Jell 107 (Products Research and Chemical Corp., Semco Division) surface treatment were fabricated in a press at 330°C under 1.7 MPa for 0.25 hr. Tensile shear strengths were determined according to ASTM D1002 using four specimens per test condition.

Composite Specimens. A 25 % solids (w/w) solution of the controlled molecular weight endcapped PAEBI prepared from 5,5'-bis[2-(4-hydroxyphenyl)benzimidazole] and 1,3-bis(4-fluorobenzoyl)benzene in DMAc was used to coat unsized AS-4 (12K tow) carbon/graphite fiber on a drum winding prepreg machine. The polymer was prepared using a 7 mole% stoichiometric excess of the activated aromatic difluoride and endcapped with 2-(4-hydroxyphenyl)benzimidazole in order to control the molecular weight and consequently improve melt flow behavior. The polymer had a T_g of 266°C and an inherent viscosity of 0.55 dL/g. The solution used to prepare the prepreg had a Brookfield viscosity of 5350 centipoise at 26 °C. The prepreg was initially dried to a tack-free state using heat lamps followed by stage-drying to 200°C in a forced air oven. The prepreg had volatile contents of ~3.5 % and calculated resin contents of 29 to 32 %. Unidirectional laminates (7.6 cm x 7.6 cm x 10-20 ply) were fabricated in a stainless steel mold by heating to 360°C under 1.4 MPa and holding for 1 hr.

The composite panels were ultrasonically scanned (C-scanned) to detect voids, cut into specimens, and tested for flexural properties according to ASTM D790 and compression properties according to ASTM D3410 Procedure B.

Oxygen Plasma Asher Exposure. Oxygen plasma exposures were performed under vacuum on thin films (1.3cm X 1.3cm, ~0.025mm thick) of the phosphine oxide-containing PAEBI in a Tegal Plasmod asher. The asher was operated at 500 millitorr, 100 Watts of radio frequency, O_2 pressure of 0.021MPa and a flow rate of 50 standard cubic centimeters per minute. Since the asher was not calibrated, simultaneous exposures of Kapton®HN film were performed. The Kapton®HN film served as a standard allowing for direct comparison and determination of relative stability of the experimental polymer. Exposures were performed up to 88 hr and the weight loss of the film was monitored as a function of exposure time by periodic removal of the sample. In each case duplicate samples were run.

Other Characterization. Inherent viscosities (η_{inh}) were obtained on 0.5% (w/v) solutions in DMAc at 25°C. Differential scanning calorimetry (DSC) was conducted on a Shimadzu DSC-50 thermal analyzer at a heating rate of 20 °C/min with the T_g taken at the inflection point of the ΔT versus temperature curve.

Results and Discussion

Isophthaloyl-containing poly(arylene ether benzimidazole) The chemical structures of the two PAEBIs evaluated in this study are presented in Figure 1.

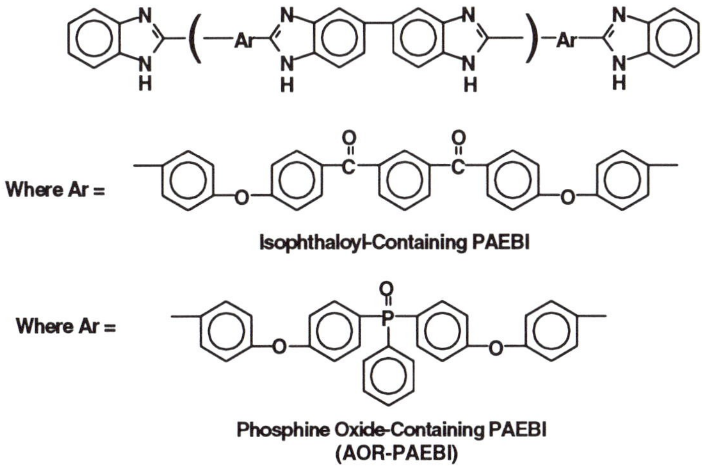

Where Ar =

Isophthaloyl-Containing PAEBI

Where Ar =

Phosphine Oxide-Containing PAEBI
(AOR-PAEBI)

Figure 1. Chemical structures of PAEBIs.

As part of an evaluation on the isophthaloyl-containing PAEBI, unoriented thin film tensile, Ti-to-Ti tensile shear and carbon fiber reinforced composite properties were determined. In order to improve melt flow, and accordingly compression moldability, controlled molecular weight and endcapped polymer was prepared at stoichiometric imbalances of 1.5 to 7 mole%. The effect of stoichiometric imbalance on T_g and unoriented thin film tensile properties is shown in Table 1. The T_g followed the expected trend of reduction with increased stoichiometric offset (i.e. reduced molecular weight). Thin film tensile properties measured at 23 and 232°C were unaffected for polymers prepared with stoichiometric imbalances up to 7 mole%. Presumably this phenomenon is a result of the relatively large molecular weight distributions exhibited by these materials when measured by gel permeation chromatography and low angle laser light scattering.[4] Moldings of the molecular weight controlled powders were prepared by heating to 365°C for 1 hr under 2.4 MPa. The moldings were subsequently machined into compact tension specimens and measured for fracture toughness. As the calculated molecular weights of the polymers decreased to ~9700 g/mole, the G_{Ic} dropped from ~1000 to ~475 J/m^2.

Table 1. Unoriented Thin Film Tensile Properties

Material Property	Stoichiometric Imbalance (mole %)				
	0	1.5	3	5	7
M_n (cal'cd) g/mole	----	46.4 k	23 k	13.7 k	9.7 k
η_{inh}, dL/g	1.99	1.65	1.42	0.84	0.55
Tg, °C (powder)	276	274	274	269	266
Tensile St., MPa 23/232°C	125/85	125/81	129/83	118/66	114/71
Tensile Mod., GPa 23/232°C	4.1/2.8	3.6/2.6	4.0/2.8	3.5/2.3	3.6/2.6
Elong., % 23/232°C	14/7	12/5	30/12	13/8	22/10

It has been demonstrated that unstressed film of this polymer was very resistant to common solvents encountered in the aircraft industry such as hydraulic fluid and jet fuel[4]. In addition, the unstressed film showed no changes in tensile properties after isothermal aging at 200°C in flowing air for 1000 hr[4].

Ti-to-Ti tensile shear properties of both unaged and aged, unstressed specimens are presented in Table 2. The specimens were fabricated at 330°C for 0.25 hr under 1.7

Table 2. Titanium Tensile Shear Strengths

Test Temp., °C	Aging Time at 200°C in Air (hr)		
	0 St., MPa	500 St., MPa	1000 St., MPa
23	22.3	23.1	21.2
177	15.4	16.8	17.7
200	13.9	17.2	16.6
232	11.4	13.3	16.1

MPa. The strengths were lower than anticipated due to adhesive failure apparently resulting from incompatibility with the surface treatment (PASA Jell 107). This material did however exhibit outstanding adhesion to copper foil and to polyimide. The exact details and results of the copper/polyimide adhesion study are unavailable for publication at this time. However, this polymer exhibited adhesion to copper far superior to that of any known polyimide. In addition, this material exhibited good adhesion to polyimide, which was a requirement for the intended application represented schematically in Figure 2. The advantages of using the isophthaloyl-containing PAEBI versus the metal process is summarized in Figure 2.

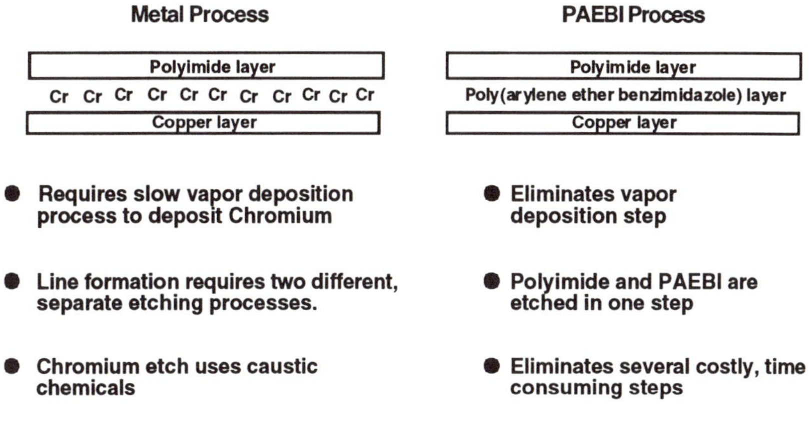

Figure 2. Comparison of process using PAEBI as an adhesion layer versus metal deposition process.

In addition to the advantages listed, the isophthaloyl-containing PAEBI exhibited other favorable properties such as initial solubility in NMP, high thermal stability, low residual stress in film, acceptable dielectric constant (3.2) and was amenable to scale-up. This material does have a relatively high moisture absorption (~4-5%) as one may expect, however, for this particular application, moisture absorption was not a problem.

Carbon fiber reinforced composite properties are presented in Table 3. The laminates were processed at 360°C for 1 hr under 1.4 MPa. The laminates exhibited high 23°C flexural and compressive properties with good retention of the flexural properties up to 232°C.

Table 3. Unidirectional Carbon Fiber (AS-4) Composite Properties

Test Temp., °C	Flex. St., MPa	Flex. Mod., GPa	Comp. St., MPa	Comp. Mod., GPa	Poissons Ratio
23	1910	135.8	1150	115.8	0.318
177	1607	102.8	----	----	----
200	1290	107.6	----	----	----
232	1124	100.0	----	----	----

Phosphine oxide-containing poly(arylene ether benzimidazole) The phosphine oxide-containing poly(arylene ether benzimidazole) (AOR-PAEBI) was prepared as previously described.[5] Unoriented thin film tensile properties of two stoichiometric offset and endcapped AOR-PAEBI are presented in Table 4. There is not much difference in the thin film tensile properties for the two offsets, however, there is a notable difference in the T_g.

Table 4. Thin Film Tensile Properties of AOR-PAEBI

% Offset	η_{inh}, dL/g	T_g.°C	Test Temp.,°C	Tensile St.,ksi (MPa)		Tensile Mod.,ksi (GPa)		Elong., %
2	0.91	376	23	18.8	(129.7)	556	(3.83)	5
			232	15.4	(106.2)	374	(2.58)	6
			316	8.0	(55.2)	275	(1.89)	14
5	0.41	364	23	19.6	(135.2)	518	(3.57)	12
			232	13.9	(95.9)	367	(2.53)	7
			316	7.9	(54.5)	229	(1.58)	24

The phosphine oxide-containing PAEBI was evaluated for stability to oxygen plasma and atomic oxygen. The exposures were carried out in an oxygen plasma asher under vacuum. Since the asher was not calibrated simultaneous exposures with Kapton® HN film were performed. Since the erosion rate of

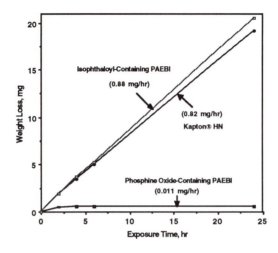

Figure 3. Oxygen plasma exposure of PAEBI, AOR-PAEBI and Kapton® HN.

Kapton® HN film by AO is well documented, it serves as a control for these exposure experiments. The Kapton® HN film was comparable in thickness to the phosphine oxide and isophthaloyl-containing PAEBI films. The results of the oxygen plasma exposures are represented graphically in Figures 3 and 4. In these experiments the films were periodically removed and weighed and the weight losses monitored over the exposure period. The AOR-PAEBI film exhibited a much lower weight loss rate than either the isophthaloyl-containing PAEBI or Kapton® HN films. It is clear that the phosphorus present in AOR-PAEBI accounts for this improved stability since the isophthaloyl-containing PAEBI, which is similar in chemical structure but contains no phosphorus, exhibits a weight loss rate comparable to that of Kapton® HN (Figure 3). A similar effect has been seen for phosphine oxide-containing polyimides[6,20], phosphine oxide-containing poly(arylene ether)s[18,19] and phosphine oxide-containing poly(arylene ether)s containing 1,3,4-oxadiazole[6], benzoxazole[20], benzothiazole[20], and quinoxaline[20] units. The phosphine oxide-containing polymers also exhibited non-linear weight loss behavior whereas Kapton® HN and polymers without the phosphine oxide group exhibited linear weight loss rates. The AOR-PAEBI exhibits an initial weight loss rate comparable to that of Kapton® HN, however after ~1-2 hr exposure in the asher the rate decreases notably. This initial weight loss appears to be the result of chemical changes taking place on the film surface which involves reaction of the phosphorus in the polymer with atomic oxygen to form a phosphate-type surface layer. This surface layer protects the underlying polymer from further exposure and subsequent erosion. This effect has been observed with other phosphorus-containing polymers such as polyphosphazenes[21,22] and phosphine oxide-containing poly(arylene ether)s.[18,19]

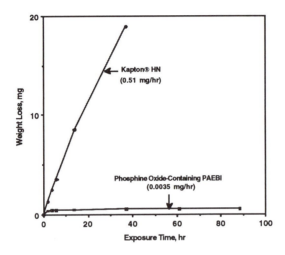

Figure 4. Oxygen plasma exposure of AOR-PAEBI and Kapton® HN for 88 hr.

To further assess the stability of AOR-PAEBI, an asher exposure lasting 88 hr was performed. The results are presented graphically in Figure 4. After ~24 hr in the asher the 0.002 in thick Kapton® HN film was completely disintegrated. The average weight loss rates for the phosphine oxide-containing PAEBI, normalized to a Kapton® HN weight loss rate of 0.82 mg/hr, are 0.011 mg/hr for the 23 hr exposure and 0.0056 mg/hr for the 88 hr exposure. The average weight loss rate of the AOR-PAEBI decreased compared to that obtained for the 23 hr exposure. This reduction in overall weight loss rate is due to the fact that for each experiment, the weight loss rate decreases after the initial 1-2 hr of exposure so that in a longer exposure the average weight loss rate will be less than that of a shorter exposure.

In order to substantiate the results obtained in the asher exposures, a more sophisticated experiment was performed.[23] This exposure was run in a continuous

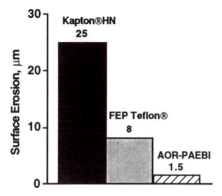

Figure 5. Continuous full duty atomic oxygen exposure of AOR-PAEBI, Kapton® HN and Teflon®. Ref. 23.

full duty fast atomic oxygen test facility located at Los Alamos National Laboratory. This facility accurately simulates the atomic oxygen present in LEO in terms of its flux (1 x 10^{17} oxygen atoms/sec·cm^2) and energy level (5eV). The total fluence for this 4 hr exposure was 0.7-1.4 X 10^{21} oxygen atoms/cm^2 which is comparable to the fluence experienced on a 40 hr space shuttle exposure. The results are represented in Figure 5. It is important to note that the graph in Figure 5 presents surface erosion not weight loss versus exposure to AO fluence. The AOR-PAEBI (1.5 μm) exhibited substantially less erosion than either Kapton® HN (25 μm = 0.001 in.) or Teflon® (8.0 μm). In addition, the AOR-PAEBI exhibited most of the erosion within the first 15 min of exposure. Extrapolation of the data obtained on AOR-PAEBI during this short term exposure to 10 and 30 years of exposure indicates that this material has sufficient promise to warrant further research and development work.

The results of the XPS analyses of AOR-PAEBI film exposed for 23 hr in the oxygen plasma asher are presented in Table 5. All photopeaks are referenced to that of carbon having a maximum taken at 285.0 eV. The control sample (unexposed) exhibited an oxygen photopeak ~532.2 eV and a phosphorus photopeak ~ 132.6 eV. The relative concentrations of oxygen and phosphorus were 17.6 and 1.2%, respectively. The control sample exhibited some minor silicon and calcium contamination, however the exposed sample did not. After exposure to the oxygen plasma the binding energies exhibited an increase to 533.2eV for oxygen and 134.6 for phosphorus. The concentration of oxygen and phosphorus increased while that of carbon decreased. In addition, the ratio of oxygen to phosphorus changed from ~15:1 to ~4:1 as a result of the exposure. The shifts in binding energies and the changes in surface concentrations of the oxygen and phosphorus

Table 5. XPS of AOR-PAEBI

Photopeak	CONTROL		EXPOSED*	
	Binding Energy, eV	Atomic Conc., %	Binding Energy, eV	Atomic Conc., %
C1s	285.0	74.3	285.0	52.9
O1s	532.2	17.6	533.2	35.6
N1s	399.9	5.6	401.3	3.1
Ca2p	347.4, 351.2	0.8	----	----
P2p	132.6	1.2	134.6	8.4
Si2p	102.1	0.5	----	----

*Exposed for 23 hr

have been observed with other phosphorus-containing polymers[6,18-22] and are indicative of the formation of a higher oxidized phosphorus species (i.e. phosphate-type). This surface layer which forms as a consequence of atomic oxygen exposure is believed to be the reason for the improved resistance that is exhibited by these materials.

Summary

Controlled molecular weight versions of the isophthaloyl-containing PAEBI exhibited high mechanical properties in the form of unoriented thin films and carbon fiber reinforced composites. In addition, this material exhibited outstanding adhesion to copper making it potentially useful in certain microelectronic applications. The phosphine oxide-containing PAEBI also exhibited high unoriented thin film tensile properties. In addition, this polymer exhibited outstanding resistance to oxygen plasma and atomic oxygen. This material is potentially useful in solar arrays, multi-layer thermal insulation blankets or coatings on spacecraft in LEO and is presently undergoing more extensive evaluation for these applications.

References

1. J. G. Smith, Jr., J. W. Connell and P. M. Hergenrother, Polym. Prepr., 32(3), 193 (1991).

2. J. G. Smith, Jr., J. W. Connell and P. M. Hergenrother, ibid, 33(1), 411 (1992).

3. J. G. Smith, Jr., J. W. Connell and P. M. Hergenrother, ibid, 34(1) 875 (1993).

4. P. M. Hergenrother, J. G. Smith, Jr. and J. W. Connell, Polymer, 34(4), 856 (1993).

5. P. M. Hergenrother, J. W. Connell and J. G. Smith, Jr., Matls. Res. Soc. Symp. Proc., Vol. 305, 21 (1993).

6. J. W. Connell, J. G. Smith, Jr. and P. M. Hergenrother, Polym. Prepr., 34(1), 525 (1993). Polymer, in press.

7. P.N. Peters, R.C. Linton and E.R. Miller, J. Geophys. Res. Lett., 10, 569 (1983).

8. D.E. Bowles and D.R. Tenney, Sampe J, 23(3), 49 (1987).

9. W.S. Slemp, B. Santos-Mason, G.F. Sykes Jr. and W.G. Witle Jr., AO Effect Measurements for Shuttle Missions STS-8 and 41-G, Vol 1, Sec 5, 1 (1985).

10. LDEF-69 Months in Space, First Post Retrieval Symposium. NASA Conference Publication 3134 Part 2. A. Levine Ed., 1991.

11. L. Leger, J. Visentine and B. Santos-Mason, Sampe Quaterly, 18(2), 48 (1987).

12. A.E. Stiegman, D.E. Brunza, M.S. Anderson, T.K. Minton, G.E. Laue and R.H. Liang, Jet Propulsion Laboratory Publication 91-10, May 1991.

13. L.J. Leger, I.K. Spikes, J.F. Kuminecz, T.J. Ballentine and J.T. Visentine, STS Flight 5, LEO Effects Experiment, AIAA-83-2631-CP (1983).

14. B.A. Banks, M.J. Mistich, S.K. Rutledge and H.K. Nahra, Proc. 18th IEEE Photovoltaic Specialists Conf.,1985.

15. K.A. Smith, Evaluation of Oxygen Interactions with Materials (EOIM), STS-8 AO Effects, AIAA-85-7021 (1985).

16. S. Packrisamy, D. Schwam, and M. Litt, Polym. Prepr., 34(2), 197 (1993).

17. P. M. Hergenrother, B. J. Jensen and S. J. Havens, Polymer, 29(2), 358 (1988).

18. C.D. Smith, H. Grubbs, H.F. Webster, A. Gungor, J.P. Wightman and J.E. McGrath, High Perf. Polym., 3(4), 211 (1991).

19. J. G. Smith, Jr., J. W. Connell and P. M. Hergenrother, Polym. Prepr., 33(2), 241 (1992). Polymer, 35(13), 2834 (1994).

20. J.W. Connell, J.G. Smith Jr. and J. L. Hedrick, Polym. Mat. Sci. and Eng. Proc., 69, 289 (1993). Polymer, in press.

21. L.L. Fewell, J. Appl. Polym. Sci., 41, 391 (1990).

22. L.L. Fewell and L. Finney, Polymer, 32(13), 393 (1991).

23. A. Shepp and R. Haghighat, SBIR Phase I Final Report NAS1-19909, 16 (1993).

RECEIVED October 24, 1994

Chapter 13

Synthesis and Characterization of Porous Polyimide Films for Dielectric Applications

O. Gain[1,2], G. Seytre[1], J. Garapon[2], J. Vallet[2], and B. Sillion[2]

[1]Laboratoire d'Etudes des Matériaux Plastiques et des Biomatériaux, Unité de Recherche Associé, Centre National de la Recherche Scientifique 507, Université Claude Bernard Lyon 1, 69621 Villeurbanne Cedex, France
[2]Unité Mixte de Recherche 102 Institut Français du Pétrol, BP 3, 69390 Vernaison, France

A methodology to prepare high temperature porous polyimide films which present a dielectric constant smaller than polyethylene ($\varepsilon' = 2.2$) has been developed. Films were obtained by casting polymers which contain thermally labile groups likely to generate gaseous products. We obtained porous polyimide films with a dielectric constant of ≈ 1.5. Transmission electron microscopy analysis of the resulting porous polyimide films showed that void size and shape depend upon the thermal treatment conditions applied. This study pointed out a qualitative relationship between volume fraction of voids and dielectric constant.

High performance polymers play an important role as packaging materials in the manufacture of microelectronic devices and components, and are finding applications as interlayer dielectrics for thin film wiring in multichip packages and chip interconnection (1-2).

Technological progress is nowadays dependent upon the breakthrough of materials possessing new or vastly enhanced properties. To develop high speed computers, a reduction in the dielectric constant of the insulating material is required in order to reduce the transmission delay time. Moreover, lower dielectric constant materials reduce crosstalk between adjacent circuit lines.

Generally speaking, polymers for electronic applications require properties such as relatively high glass transition temperatures, good processability and planarization ability, low dielectric constant, and low thermal expansion coefficient.

Polyimides are very good candidates to meet all of these requirements. A great amount of synthetic work has been devoted to prepare polymeric materials possessing a dielectric constant below 3.0 and a glass transition temperature in the range of 280 - 350°C. Thus, important achievements have been realized by th incorporation of fluorine atoms, mainly through the hexafluoroisopropylidene group, which further minimizes moisture absorption (3). Similar results are obtained by the incorporation of bulky substituents (4), or by blending with oligomeric additives (5). A dielectric constant of 2.5 seems to be the ultimate value that is achievable through such structural modifications owing to the intrinsic polar nature of any imide units(6).

In all these studies, the reduction of the dielectric constant was achieved through a decrease in chain-chain electronic interactions related to steric hindrance.
The authors pointed out the importance of free volume in the polymer, inducing the development of other systems where the dielectric constant is reduced by generating voids in the polymer to bring the constant closer to the value of air *(7-9)*.

Porous polymers of various void size - from nano to macro scale- can be obtained depending upon the methodology employed. Voids can be achieved by, namely :
- the decomposition of a foaming agent either by thermal *(10-12)* or plasma treatment *(13)*,
- the processing of organic *(14)* and inorganic aerogel *(15)* (sol-gel processing),
- the paraffin oil dispersion methodology *(16)*,
- the purge with nitrogen of an aqueous precursor system *(17)*.

Applications of such foams range from low cost domestic foams to high cost selective membranes and filters. Their use for electronic applications are considered by *J. Hedrick et al (10)* and *Mukherjee, Shyama (15)*. The *Mukherjee, Shyama* study deals with polymer blends between a Xydar matrix and inorganic or inorganic-organic silicate aerogels. The Xydar polyester presents a dielectric constant of 2.72, while the blend with organic aerogel exhibits a value of 2.46.
J. Hedrick et al. work on block copolymers of the A-B-A type is based on polyphenylquinoxaline as the matrix and polyoxypropylene oxide as the thermally labile block. By adjusting the molecular weights of the two components, the volume fraction and size of the resultant voids can be varied. The resulting PPQ foams obtained, having an average cross section from 8 - 10 nm, show a dielectric constant of ≈ 2.4 while current PPQ exibits a value of 2.8.

We report in this paper, the realization of porous polyimide films obtained by degradation of carbonate groups. Firstly, a study on model compounds allows us to determine the effect of the structure on the thermolysis, secondly the synthesis of modified polyimides is conducted and the dielectric characterization of the porous films obtained is achieved. As a matter of fact, such a system is expected to lead to a foam structure which also exhibits good mechanical properties because the foaming process does not cleave the polymer chain.
The choice of the polymer takes into account two important parameters for the generation of a foam :
- solubility in a volatile solvent allowing its complete removal after drying at low temperature,
- high value of the glass transition temperature preventing the expanded film from collapsing.

Experimental

Reactants :
- BTDA : 3,4,3',4'-Benzophenone tetracarboxylic dianhydride was purchased from Allco
- BHTDA : 3,4,3',4'-Benzhydrol tetracarboxylic dianhydride was synthesized in our laboratory
- MDA : Methylene diamine was provided by BASF

- A6F : 4,4'-[2,2,2-trifluoro -1-(trifluoromethyl) ethyldene]bis-benzenamine was purchased from CHRISKEV Company
- TBDC : tertio-Butyl dicarbonate was purchased from ALDRICH
- DMAP : 4-Dimethylamino pyridine was purchased from FLUKA
- γ–BL : γ–Butyrolactone
- NMP : N-methylpyrrolidone
- DMAc : N,N-dimethylacetamide

Measurements :
→ Dielectric measurements were carried out with a Dielectric Analyser 2970 TA Instrument on large temperature (30-350°C) and frequency (1 Hz-200 kHz) ranges with sputter coated or single surface sensors for characterization of films or solutions respectively.
→ Thermogravimetric analysis (TGA) measurements were performed on a Netzch TG209 cell and DuPont TGA2950 under argon atmosphere.
→ ^1H Nuclear Magnetic Resonance (NMR) was performed on a Brucker AC 250 MHz (5.8T) spectrometer in DMSO-d6 at 25°C .
→ Scanning Electron Microscopy (SEM) was realized on a Jeol 1840A.

Presentation of the model compounds

We used the labile t-butyloxycarbonyl group, which is well-known as protecting/deprotecting agent able to undergo thermolysis around 150°C *(18)*.
    In order to study structure/thermolysis relationship, we have prepared different diimide compounds.

Synthesis.   The reaction of benzhydrol tetracarboxylic dianhydride with 4-t-butyl aniline and 3-phenoxy aniline in NMP leads to the model compounds presented figure1.

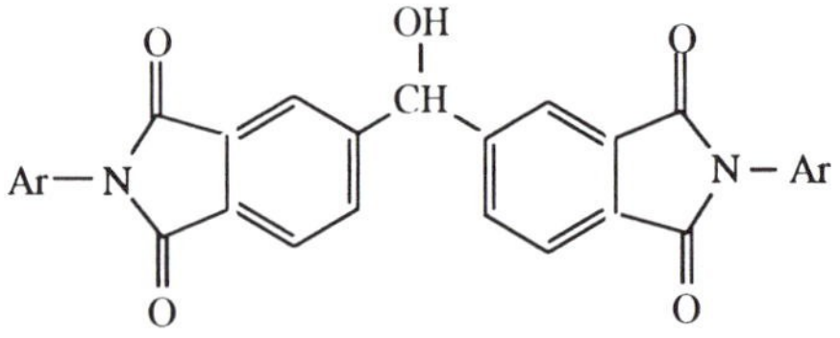

Ar : model 1-> 4tBu-C_6H_4-   model 2-> 3PhO-C_6H_4-

Figure 1. Benzhydrolimide models.

The compoud 1 is crystalline (T_{melt}=278°C) and the substrate 2 is an amorphous compound.

To achieve carbonatation, we employed t-butyl dicarbonate (TBDC). The reaction occurs at 25°C in γ-butyrolactone, with a small amount of dimethylamino pyridine, as described in figure 2.

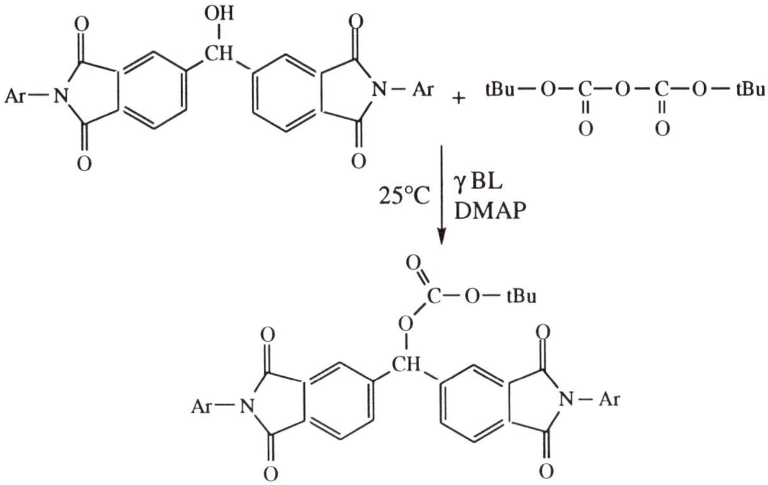

Ar : 4 tBu-C_6H_4-   ->   carboante 3
3 Ph-O-C_6H_4-   ->   carbonate 4

Figure 2. Carbonatation of model compounds

Thermal behaviour. Thermal decomposition of the side chain t-BOC groups was investigated by TGA and DSC for carboante 3 and 4.
The results are summarized in Table 1.

Table 1. Thermal behaviour of model compounds.

	Carbonate 3	Carbonate 4
Molecular weight	685 g/mol crystalline	758 g/mol amorphous (Tg≈70°C)
TGA		
loss at 220°C	14,9 %	15,3 %
theoretical loss	14,5 %	13,9 %
DSC		
endotherm	220°C	187°C
exotherm	226°C	
endotherm	280°C	

In TGA, the compounds are found to be stable up to 130°C, but above 150°C the t-BOC group undergoes a rapid thermolysis by releasing carbon dioxide and isobutene.
DSC thermograms reveal an endothermic event corresponding to the thermolysis of t-BOC group at 187°C (carbonate 4) and 220°C (carbonate 3).This difference is presumably ascribable to the difference of physical state (crystalline vs amorphous).
The exothermic and endothermic phenomena at 226°C and 280°C are due to crystallization and melting of the compound 1.
DSC analysis points out the influence of the chemical structure on the temperature degradation of t-BOC groups of the benzhydrol diimide formed.

Polymer Formation

Polymers were prepared by a one-step polycondensation reaction between dianhydride and diamine. In order to employ a dianhydride with a pure benzhydrol structure, benzhydroltetracarboxylic dianhydride (BHTDA) was prepared by reduction of benzophenone tetracarboxylic dianhydride (BTDA) with potassium borohydride. This selective reagent toward the ketonic function avoids further reduction leading to methylene groups as we observed in our laboratory using a conventional catalytic (Pd/C) hydrogenation.

Due to the solvent retention of BHTDA evaluated by NMR around 8%, we realized polycondensations with an excess of dianhydride (0 to 10%) in order to obtain an inherent viscosity in the range of 0.4 dl/g high enough to get films with good properties.

Polybenzhydrolimide precursors.

Synthesis. The BHTDA and the diamines were condensed in an m-cresol solution (20 % w/w) at 180°C during 3 hours (Figure 3). Then the polyimide was precipitated in methanol, crushed, washed and dried at 140°C in vacuum during a few hours.

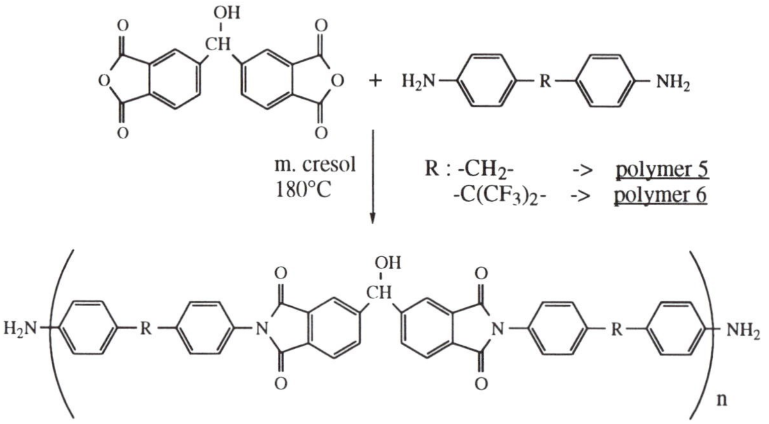

Figure 3. Synthesis of polyimide precursor

Physical and chemical characterization.   The glass transition measured by DSC and the inherent viscosity is reported in table 2 with BHTDA/diamine ratio.

Table 2. Characterization of precursors.

	Tg (°C)	η_h (dl/g)	ratio
Polymer 5 BHTDA/MDA	273	0.45	1.06
Polymer 6 BHTDA/A6F	294	0.35	1.08

NMR analysis realized on polymer 5 shows for ^{1}H spectrum two doublets at 6.5 and 6.9 ppm and for ^{13}C three peaks at 114, 142.6 and 147 ppm attributed to the terminal aromatic methylene aniline protons. This assignement is confirmed by NMR spectrum of a BHTDA/2MDA model isolated pure(figure 3 with n=1).

Polybenzhydrolimide carbonate.

Synthesis. Polybenzhydrolimide was converted to the corresponding polybenzhydrolimide carbonate at ambient temperature by treatment with an excess of DBDC and a catalytic amount of DMAP in γ–butyrolactone (Figure 4).

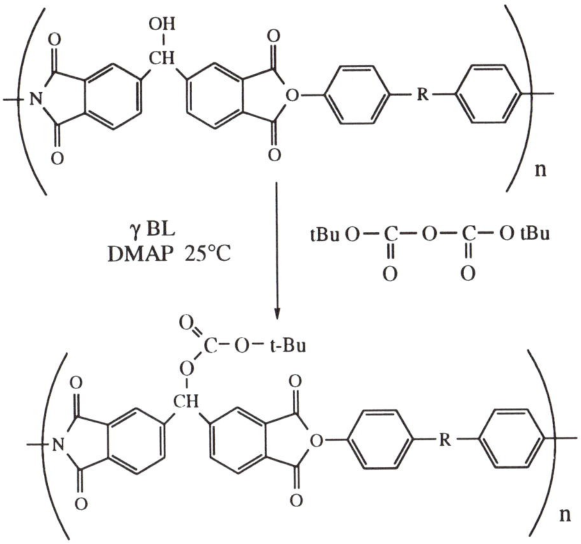

R : -CH₂- -> polymer 7
-C(CF₃)₂- -> polymer 8

Figure 4. Synthesis of t-BOC modified polymer.

We observed that the inherent viscosity of the carbonated polyimide increased with the TBDC/benzhydrolimide molar ratio as it is indicated in table 3, in the case of polymer 5.

Table 3. Polymer 5 : Molar ratio versus inherent viscosity.

molar ratio	viscosity (dl/g)
1	0.49
1.3	0.48
1.4	0.98
1.5	1.2

We reasoned that this increase in viscosity was due to a chain growing mechanism through the reaction of the terminal amino group with the plausible TBDC. In order to confirm this hypothesis, we checked for different plausible side reactions between the organic functions present in the medium.

Side reactions :
- reaction of benzhydrol t-BOC carbonate with aniline
-> *no reaction.*

- reaction of primary amine on TBDC with and without DMAP.

> *we obtain carbamate and/or urea and its derivaties.*

- reaction of carbamate group toward amine and TBDC.

> *no reaction.*

- reaction of TBDC on urea.   > *formation of t-BOC derivatives.*

These results clearly show :
    - the lack of reactivity of the benzhydrol t-Boc carbonate toward amine. So the increase of viscosity observed with an excess of TBDC is not due to crosslinking.
    - the unreactivity of carbamate toward amine and TBDC.
    - the reactivity of TBDC toward amine and urea.

These results are consistent with a chain growing phenomenon by the coupling of amine end groups and the formation of urea links. A plausible mechanism is shown in figure 5.

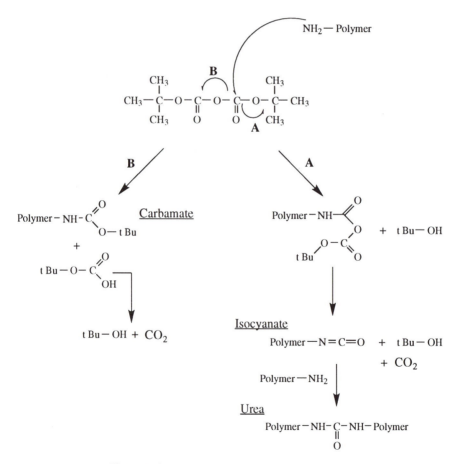

Figure 5. Scheme of chain growing mechanism.

Physical and chemical characterization.   As expected according to the model studies, the thermal degradation of t-BOC group occured around 150°C (TGA analysis).

The content of t-BOC groups evaluated by TGA and NMR was 78 % for polymer 5.

With the 6F polybenzhydrolimide carbonated, we encountered the same phenomenon as for polymer 5. The difference is the higher glass transition of polymer 6 (table 2).

Formation of Porous Films.

This study was developed with polymer 5.

Films were prepared by casting a DMAc solution on a glass plate. They were thermally dried at 100°C for 3 hours. Afterwards, they were removed from the glass plate and put in an oven under vacuum at 100°C for 2 hours.

Realization of foam structure.   The thermolysis could be obtained either using different *ramp* or by *isotherm* (flash cure).

Using TGA we simulated different cure conditions. The thermogravimetric analysis allowed us firstly to follow the decomposition of the t-BOC under varying heating conditions , and secondly to carry out a visual inspection of the film after curing.

As a matter of fact, visual examination of the sample gave a good information about the "structure" : an opaque aspect means that the foam structure remained and that the polymer had not collapsed.

Theoretically the aspect of the sample is conditioned by both the shape of the particles and the difference between the refractive index of the two phases *(19)*. The scattering losses will be minimized only for composites with particle domains smaller than the wavelength of light. Thus we think that opaque samples have voids with dimension in the visual wavelength range (up to 400nm).

With this method, critical conditions have been found under which the samples remained transparent and above which they were opaque. These conditions were 20°C/min for the ramp and 180°C for flash cure, for a thickness of ≈ 50μm.

We must emphasize that these critical conditions are dependent upon the sample thickness.

S.E.M. analysis for flash cure conditions.   Films with 50 μm and 30 μm thickness have been submitted to different flash cure and studied by SEM to get information on the size scale of pores.

Figures 6-8 show sections of polmer-5 for different flash cure conditions. It is evident that a porous structure is obtained with pores having an average cross section from 0.2 - 3 μm. After flash curing, we observed only a few pores scattered in the matrix (figure 6) or a foam network more or less dense(figure 7-8).

Furthermore, we see that the degradation of t-BOC groups induces an expansion of the sample thickness up to 120 μm.

For thinner films (30μm), we have obtained, after a flash cure at 155°C, a foam with pores having a cross section from 5 to 40μm (figure 9) with a thickness of 85μm.

These figures indicate that flash cure conditions depend also upon the thickness and that the distribution of size is a function of the flash cure condition and the thickness of the sample.

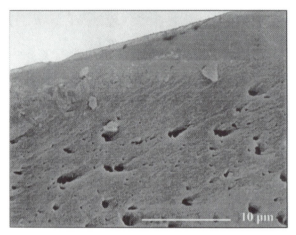

Figure 6. **Foam 1** Magnification X3500; Void 0.4-2μm

Initial thickness 50 μm; After thermal treatment 50 μm
Permittivity 2.9; Cure condition 200°C for 4 mn.

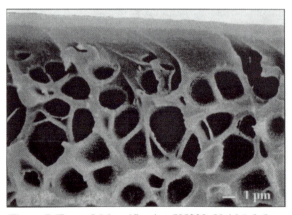

Figure 7. **Foam 2** Magnification X5000; Void 1-3.5μm

Initial thickness 50 μm; After thermal treatment 87 μm
Permittivity 1.8; Cure condition 245°C for 3 mn.

Figure 8. **Foam 3** Magnification X5000; Void 0.2-1μm

Initial thickness 50 μm; After thermal treatment 120 μm
Permittivity 1.5; Cure condition 255°C for 3 mn.

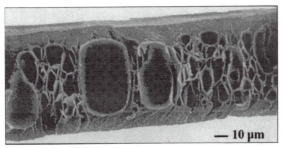

Figure 9. **Foam 4** Magnification X500; Void 5-40μm

Initial thickness 30 μm; After thermal treatment 85 μm
Permittivity 2.2; Cure condition 155°C for 5 mn.

Dielectric Characterization.

Measurements on solutions. They were performed on interdigitated sensors with a solution at 30% in dioxane (figure 10).

We obtained two types of information : first a strong decrease of ε' related to the removal of solvent, then, during an heating ramp from 110°C to 300°C, a sheer decrease around 150°C for the modified polymer (polymer 8).

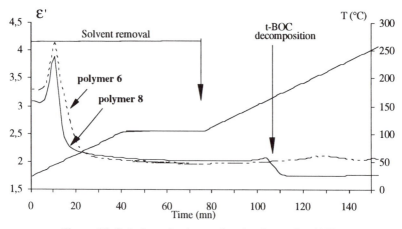

Figure 10. Solution of <u>polymer 6</u> and <u>polymer 8</u> at 1kHz.

The variation of ε' around 150°C was not observed for the precursor (<u>polymer 6</u>). So we can attribute the variation of ε' at this temperature to the decomposition of t-BOC groups. These qualitative results allowed us to follow "in situ" variations of ε' and we think that this method could be a useful process for in situ cure monitoring of foam structures.

Measurements on films.   The evolution of the dielectric constant for <u>polymer 5</u> is presented at different stage figure 11.

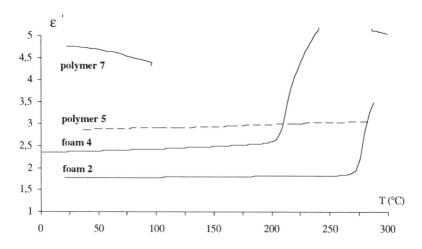

Figure 11. Evolution of the dielectric constant for <u>polymer 5</u> at 1kHz

On these dielectric spectra, we observed an increase of the dielectric constant around 275°C, for <u>foam 2</u> and <u>polymer 5</u>, ascribable to the glass transition of such polybenzhydrolimide structure as observe previously by DSC.

The variation of ε' at 200°C for <u>foam 4</u> is due to a modification of the sample thickness during the measurement. This foam possess "hudge" voids which induce poor thermomechanical properties. So, the force applied by the electrodes on the sample collapses the structure at that temperature. The result is an increase of the capacitance caused by the decrease of the thickness.

Dissipation factor (tan δ) spectra for precursor and porous films point out a sub-glass relaxation around 150°C (figure 12) attributed to local motion of the imide linkage.

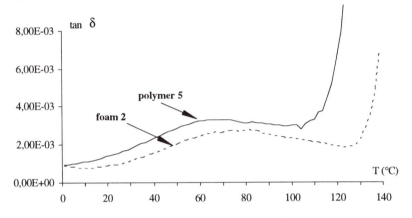

Figure 12. Dissipation factor of polymer 5 at 1kHz.

The SEM study of film sections enabled us to point out the relationship between foam structure and dielectric constant as we can see in table 4.

Table 4. Evolution of permittivity for polymer 5 as a function of the morphology.

	void size (μm)	Void observation		permittivity (100°C, 1kHz)
		Hole	Distribution	
precursor	--	--		2.9
foam 1	0.4 - 2	small	scattered	2.9
foam 2	1 - 3.5	small	dense	1.8
foam 3	0.2 - 1	very small	very dense	1.5
foam 4	5 - 40	large	dense	2.2

Obviously there is a relationship between morphology and concentration of voids and permittivity. But, we have not found an accurate method which can determine the "cell volume", so we cannot say if it is a linear dependence, an inverse volume fraction model, or another one.

Similar results were obtained with the polymer 6, see table 5.

Table 5. Dielectric characteristics of polymer 6.

	void size (μm)	permittivity (100°C, 1kHz)	tan δ (100°C, 1kHz)
precursor	--	3.1	$1\ 10^{-2}$
foam	8 - 50	2.2	$4\ 10^{-3}$

Foam formation.

Figure 13 represents the dependence between flash cure condition, voids and permittivity for polymer 5.

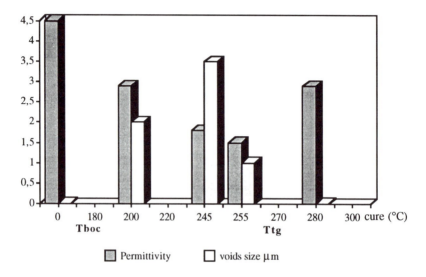

Figure 13. Dependence of voids size and permittivity versus flash cure condition.

T_{boc} corresponds to the beginning of t-BOC thermolysis (180°C in our exemple) and T_{tg} is the glass transition of the end product (270°C). These two values are characteristic of the polymer.

Below T_{boc}, no modification of the structure occurs and the hight permittivity is mainly due to the carbonate polar groups.

Above T_{tg}, chain motion appears leading to collapsed structure and the permittivity measured is the "normal" polybenzhydrolimide value.

Between these limits, foam structures were realized whose permittivity depends upon void size and concentration. We think that these parameters must be related to the viscosity of the medium; Thus, around T_{boc} a quite low viscosity allows chain motion and the structure tends to collapse leading to a poor concentratioin of voids. Near T_{tg}, a high viscosity hinders any motions and the structure presents numerous voids.

Conclusion.

This study shows that porous submicron films can be obtained by decomposition of t-BOC pendent groups .
Compared to other methods : incorporation of fluorine groups, bulky substituents, and flexible links; the formation of foam appears to be the best one to achieve low values

for the dielectric constant. In the case of our polybenzhydrolimide, this method allows a 40% decrease in ε' (2.9 -> 1.5).

The more critical parameters appear to be the glass transition of the modified and unmodified polymers which govern the collapsing effect. Flash cure condition is the more effective process to obtain porous films.

Our goal is to quantify the volume fraction of voids in order to establish a model calculation of the dielectric constant.

References

1   Hergenrother P. M., *Polymer Journal*, **1987**, *19*, 1,pp73.
2   Sillion B., *Bull. Soc. Chim. France*, **1987**, *4*, pp696.
3   St. Clair A. K., St. Clair T. L. and Winfree W. P, *Polym. Mat. Sci. and Eng.*, **1988**, *59*, pp28-32.
4   Ichino T., Sasaki S., Matsuura T. and Nishi S., *J. of Polymer Science Part A: Polym. Chem.*, **1990**, *28*, pp323-331.
    Ichino T., Sasaki S., *J. of Photopolymer Sciencee and Technology*, **1989**, *2*, 1, pp39-.
5   Stoakley D. M. and St. Clair A. K., *Polym. Mat. Sci. and Eng.*, **1988**, *59*, pp33-36.
6   Mercer F. W. and Goodman T. D., *Polym. Prep. Am. Chem. Soc, Div. Polym. Chem.*, **1991**, *32*, 2, pp189-190.
7   Gain O., Boiteux G., Seytre G., Garapon J., Sillion B., *IEE, Conf. Publ.*, Manchester, UK, **1992**, *363*, pp401-404.
8   Misra A. C., Tesoro G., Hougham G. *Polym. Prep. Am. Chem. Soc., Div. Polym. Chem.*, **1991**, *32*, 2, pp191-192.
    Misra A. C., Tesoro G., Hougham G. and Pendharkar S. M., *Polymer*, **1992**, *33*, 5, pp1078-1082.
9   Min Sheng Wang, Jin Weifang and Liu Ziyu, *Proceeding of the 3rd International Conference on Properties and Applications of Dielectric Materials*, Tokyo, JP, **1991**, pp77-80.
10 Hedrick J., Labadie J., Russell T., Wakharkar V., Hofer D., "New synthetic copolymers for low dielectric high temperature polymer nanofoams" *in Advance in Polyimide Science and Technology*, ed C. Feger, M. M. Khojasteh, M. S. Htoo, Technomic, Lancaster,1993, pp184-197.
11 Memeger W. Jr. (E.I. Du Pont de Nemours and Co.), *US Patent n°4 178 419*, **12.11.79.**
12 Memeger W. Jr. (E.I. Du Pont de Nemours and Co.), *US Patent n°4 226 949*, **10.07.80.**
13 Loy D. A., Buss R. J., Assink R. A., *Polym. Prep. Am. Chem. Soc., Div. Polym. Chem.*, **1993**, *34*, 1, pp244-245.
14 Pekala R. W., Schaefer D. W., *Macromolecules*, **1993**, *26*, pp5487-5493.
15 Mukherjee S. P. and Wang D. W. (International Business Machines Corp.),*Eaur. Pat. Appl.*, EP 0 512 401 A2.
16 Sherrington D. C. , *Makromol. Chem, Macromol Symp*, **1993**, 70/71, pp303-314.
17 Palani Raj W. R., Sasthav M., Cheung H. M., *J. of Applied Polymer Science*, **1993**, 49, pp1453-1470.
18 Kwang-duk Ahn, Young-Hun Lee, Deok-II Koo, *Polymer*, **1992**, *33*, 22, pp4851-4856.
19 Nowak B. M., *Adv. Mater.*, **1993**, *5*, 6, pp422-433.

RECEIVED February 10, 1995

Chapter 14

Synthesis and Characterization of Melt-Processible, High-Molecular-Weight Poly(amide-imides)

V. N. Sekharipuram, I-Yuan Wan, S. S. Joardar, T. C. Ward, and J. E. McGrath

Department of Chemistry and National Science Foundation Science and Technology Center, High Performance Polymeric Adhesives and Composites, Virginia Polytechnic Institute and State University, Blacksburg, VA 24061-0344

The generation of soluble, melt-processible, fully cyclized and wholly aromatic poly(amide-imides) is demonstrated. This has been afforded by first synthesizing controlled molecular weight polymers by the polycondensation of bis(trimellitoyl) oxydianiline [BTM-ODA] with 4,4' oxydianiline [ODA] using benzoic acid or t-butyl benzoic acid as the monofunctional end capper under Yamazaki reaction conditions. The complex processes occurring during the molding of these poly(amide-imides) were investigated using a variety of techniques. The experimental results show that the homopolymers were able to crystallize when heated to 50~75°C above T_g. Extensive characterization afforded a better insight into the thermal transitions that occur at such molding temperatures. The crystallinity in these systems was disrupted by the generation of co-poly(amide-imides) by the reaction of BTM-ODA with 1,3-phenylene diamine. The amorphous polymers could now be molded into transparent, tough plaques under molding conditions at about 50~75°C above the T_g.

Aromatic poly(amide-imides) have attracted considerable attention in the development of heat stable, high performance polymers since they have been first reported in the early seventies (1). These polymers have excellent high temperature properties with glass transition temperatures (T_g) typically above 270°C. Torlon[TM], one important commercially available poly(amide-imide), is considered to be a reaction product of trimellitic acid chloride and a mixture of the aromatic diamines. This reaction is believed to be conducted by the classical "two step" procedure - the first step involving the synthesis of the intermediate poly(amide-amic acid), followed by a high temperature imidization.

The conditions for the commercial synthesis have not been fully disclosed in the literature. However, it is believed that the second step is often halted at some intermediate stage of imidization and molecular weight growth and the reaction may only attain completion during the molding process. Thus, it is not an entirely

0097–6156/95/0603–0214$12.00/0

unexpected feature that the melt viscosity profile of poly(amide-imides) during molding is often complex due to chain extension and thermal cyclization. It is well known that the thermal imidization is often associated with an infusible product and results in difficulty in processing. This could very well perhaps account for the poor melt flow and processible behavior often associated with aromatic poly(amide-imides) in general. This often necessitates maintaining the polymer system for extended periods of time at temperatures above the T_g during molding in order to ensure optimum physical behavior. It has been extensively reported in the literature for polyimide systems that the kinetics and mechanisms of thermal imidization processes, particularly in the solid phase, are very complex *(2-4)*. Very often, the products have poorly defined molecular architectures such as uncontrolled molecular weights and end groups and possibly, the formation of branches and crosslinks *(3,5-7)*. This would lead to high chain rigidity resulting in rather undesirable properties including loss of thermoplastic processability.

These problems have been effectively addressed in our laboratories by the use of synthetic techniques that permit the introduction of molecular weight and end group control *(8-11)*. It is therefore anticipated that the development of well defined molecular architecture coupled with control over molecular weight and end groups would enhance existing advantages of aromatic poly(amide-imides) while overriding their current drawbacks. The current paper focuses on the thermal transition behavior in the all-para, aromatic poly(amide-imides) generated by the incorporation of the 4,4'-oxy-bis(phenylene) spacers between the amide- and imide-links in the repeat unit. The intent was to develop a better understanding of the complexity of the processes that need to be taken into account during the molding operations. The molecular structure of the poly(amide-imide) system is shown below :

R = H, t-butyl

Experimental work

Synthesis and Characterization of Poly(amide-imides). The synthesis was effected by the solution condensation of the bis(trimellitimide) of *4,4'-* oxydianiline **[BTM-ODA]** with *4,4'-* oxydianiline **[ODA]** under catalytic conditions stipulated by the Yamazaki reaction *(12-14)*. This involves the complexation of the dicarboxylic acid with triphenyl phosphite **[TPP]** and pyridine **[Py]** in a solvent system consisting of **NMP** containing **LiBr**. The role of halide salts such as **LiBr** is to ensure that the poly amide formed is maintained in solution as, the homogeneity of the reaction medium has been found to be critical in the polycondensation reaction. This approach to poly(amide-imide) synthesis was adopted by Yang *et al. (15-19)* wherein a variety of diamines were reacted with a series of aromatic dicarboxylic acids that contained preformed imide moieties. In case of the polymer systems generated in our laboratories for the current study, molecular weight was controlled by stoichiometric offset of the monomeric reactants. Non-reactive chain ends were effected by employing either benzoic acid **[BA]** or *4- tert.* butyl benzoic acid **[TBA]** as the monofunctional end capper. The *t-*

butyl groups could be analyzed by proton NMR and utilized as one measure of the number average molecular weight.

The synthesis of **BTM-ODA** followed the procedure recommended by Wrasidlo and Augl *(20)* which has been appropriately modified by employing solution imidization techniques developed and optimized in our laboratory *(8-11, 21-24)*. **ODA** of fairly high purity (>99.5%, mp = 191.5°C) is available from Chriskev Co. as a slightly tan crystalline powder. As a monomer for polymerization reactions, a much higher purity is called for and so **ODA** was sublimed under a high vacuum at about 185~190°C to yield a white material that was then used in all reactions. **BA** and **TBA** were procured from Aldrich Chemical Co. and purified by recrystallization from a solution in acetone/water mixture, dried under vacuum and stored in a dessicator.

The polymers were synthesized by the following procedure: Lithium Bromide, **LiBr** (1.6 g.) was charged into a four-necked reaction flask that was fitted with a nitrogen adapter, a thermometer and a reflux water-condenser. The assembly was flame-dried under a steady stream of dry nitrogen to remove moisture that could be present adsorbed to the walls of the glass vessel or present due to the hygroscopic nature of LiBr. After cooling, the diimide acid, **BTM-ODA** (0.01 mols, 5.404 g.) was charged into the reaction flask along with a stoichiometrically equivalent amount of **ODA** (0.01 mols, 2.002 g.). In case of very high, uncontrolled molecular weight systems, the molar ratio was 1:1. Where controlled molecular weight polymers were being generated, a stoichiometrically offset quantity of **ODA** along with an appropriate amount of the monofunctional end-capper were added. The entire reaction mixture was dissolved in dry, purified **NMP** at 15~20 wt. % solids. It was observed that **BTM-ODA** shows relatively poor solubility in **NMP** at room temperature. Triphenyl phosphite, 0.02 mols along with 0.02 mols pyridine was now introduced into the system. The entire assembly was immersed in a silicone oil bath that was heated to about 115°C. As the reaction mixture temperature commenced to rise, gradual dissolution of the **BTM-ODA** was observed. The reaction temperature was allowed to stabilize at about 100°C and held there during the entire time of the reaction. After about 1 1/2 hours, the viscosity of the reaction solution commenced to rise. The uncontrolled molecular weight polymer experiments showed very high viscosities at the end of 3 hours. Therefore, the typical reaction times were standardized at 3 hours for all systems. Once the reaction was stopped, the polymer solution was cooled and added into a 5-10 fold excess of methanol as a thin stream agitated in a high speed blender. A fibrous polymer precipitated, which was filtered, dried and redissolved in **DMAC**. The entire process was repeated. and the purified polymer was dried in a vacuum oven at 175°C overnight. The synthesis of controlled molecular weight poly(amide-imides) using the catalytic amidation is described in **Scheme 1**. Good control of the target molecular weights was achieved and the details have been reported elsewhere *(25)*. The molecular weight characteristics of the polymers has been summarized in **Table I**.

Thermal Analysis. Differential Scanning Calorimetry was performed using V 4.0 B Dupont 2100 instrument. Scans were run at 10°C/min and values were obtained from both, the first as well as the second heat. Dynamic thermogravimetric scans were performed employing V 4.0 D Dupont 2100 instrument. Scans were run in air at a rate of 10°C/min and the temperature reported is for 5% weight loss.

Dynamic Mechanical Analysis. The dynamic mechanical thermal scans were run on the Polymer Laboratories DMTA at a heating rate of 2°C/min and at a frequency of 1 Hz in the bending mode.

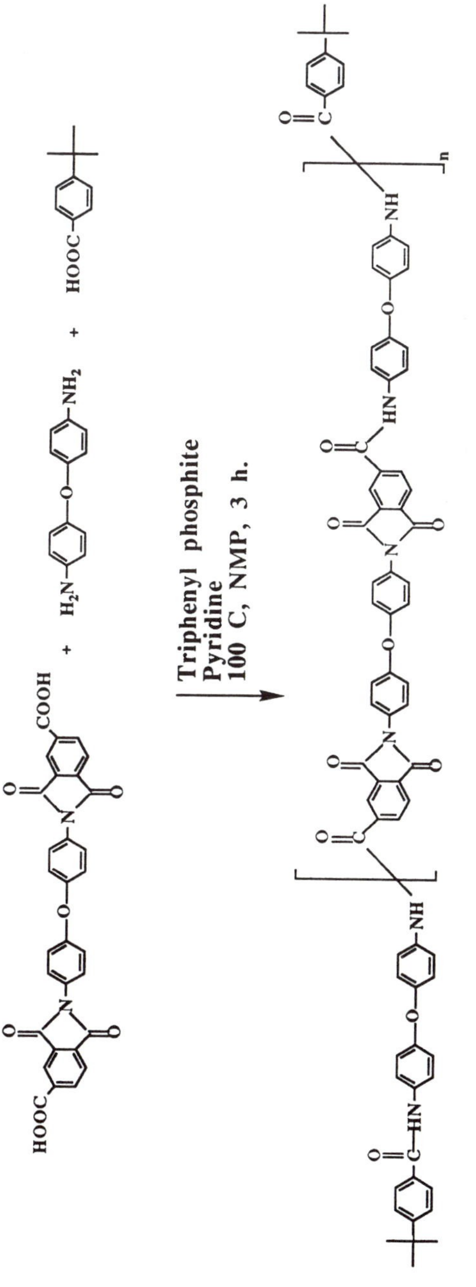

Scheme 1. Synthesis of controlled molecular weight poly(amide-imides) by the application of the Yamazaki Reaction.

Table I. Summary of molecular weight characteristics of poly(amide-imides) determined by a variety of methods

Polymer	$[\eta]^{25C}$ (dl/g)[a]	$[\eta]^{25C}$ (dl/g)[b]	VPO $<M_n>$ (Kg/mol)	LS $<M_w>$ (Kg/mol)	$[\eta]^{60C}$ (dl/g)	GPC $<M_n>$ (Kg/mol)	$<M_w>$ (Kg/mol)	$<M_w>/<M_n>$
A	2.500							
B	0.58	0.50	8.9		0.48	8.2	15.9	1.93
C	0.61	0.55	10.1	20.1	0.54	10.3	20.5	1.99
D	0.75	0.69		32.0	0.72	14.5	31.2	2.15
E	0.83	0.74		34.0	0.71	16.6	37.7	2.27
F	0.91	0.79		40.4	0.74	12.2	43.1	3.53
G	1.08	1.29	17.2	61.5	1.08	18.5	49.9	2.70

a Measurements made in NMP
b Measurements made in DMAC

Wide Angle X-Ray Scattering (WAXS). WAXS measurements were conducted by using a Scintag XDS 2000 X-ray diffractometer of Cu Kα radiation with 0.154 nm wavelength. The scanned angle, 2θ, was from 10° to 30° at a scan rate of 0.5°/min.

Results and Discussions

The thermal characterization data obtained demonstrated 5% weight loss in air by dynamic thermogravimetric analysis in all cases around 450°C. A representative TGA thermogram is reproduced below in **Figure 1**. The glass transition temperatures of the **ODA**-based poly(amide-imides) measured by DSC were observed to be about 275°C. Early attempts to melt process these polymers at about 50°C above the T_g resulted in extensive degradation and very little melt flow, even though the materials were soluble in a variety of solvents. The products of the attempted molding showed poor homogenization after an initial onset of flow. This confirmed an early suspicion that a more complex set of solid state transitions exist that cannot be accounted for by simple amorphous polymers.

A DSC thermogram of a polymer sample heated well above its T_g but below its decomposition temperature in order to detect other thermal transitions is shown in **Figure 2**. An apparent melt endotherm peak is manifested at about 419°C. A DSC scan run after annealing the polymer in the hot press at 325°C for 10 minutes is shown in **Figure 3**. The endotherm corresponding to the glass transition at ~275°C has virtually disappeared which was further indicative of the crystallization process. Thus, while the initial worked-up polymer shows very little crystallinity, as is evidenced by the ready solubility of the polymer, thermal treatment appears to enhance order. This has now been verified by wide angle X-ray scattering studies (WAXS) of samples subjected to a variety of thermal treatments. **Figure 4** is the WAXS diffractogram of the initial "as made" polymer which can be seen to be essentially amorphous.

The polymer was then subjected to annealing in the hot press for different times varying from 3-30 minutes. It was noticed that significant crystallinity develops even after 3 minutes of annealing, as is seen by a distinct sharpening of the diffraction pattern. Annealing for 5 minutes causes a marginal increase in the crystallinity. No further increase in crystallinity is observed even with annealing treatment for 30 minutes. **Figure 5** demonstrates the effect of annealing for 3 & 5 minutes.

A sample of crystallized polymer was heated above the T_m to about 425°C and rapidly quenched to confirm that indeed crystallization was occurring. Examination of this sample by WAXS again revealed essentially an amorphous halo as presented in **Figure 6**.

These results have been further confirmed by Dynamic Mechanical Thermal Analysis of solution cast films of poly(amide-imides). **Figure 7** shows a comparison of tan δ profiles of films scanned before and after annealing at 325°C for 60 minutes. The scans were made at a frequency of 1 Hz. and the heating rate was 2°C/min.

It is quite evident that the annealing process shifts the tan δ peak from 282°C to 302°C. Moreover, a reduction in the size of the tan δ peak is observed. This is further evidence for the crystallization process. Though not displayed in **Figure 7**, the storage modulus of the samples show a typical drop in value corresponding to the melting process after showing a plateau in the rubber-elastic regime.

Subsequent attempts to redevelop crystallinity in quenched samples were not successful. An examination of the dynamic mechanical thermogram in **Figure 7**

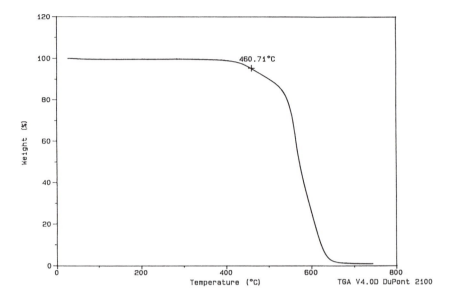

Figure 1. Dynamic thermogravimetric scan of a representative poly(amide-imide) homopolymer synthesized by the "Yamazaki" Route in air at a heating rate of 10°C/minute.

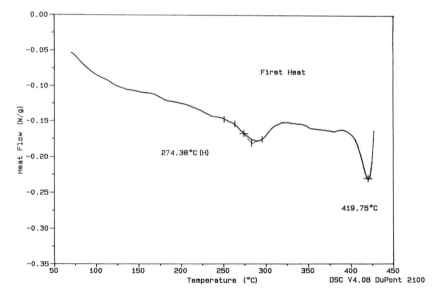

Figure 2. The first heat Differential Scanning Calorimetry thermogram of a representative poly(amide-imide) homopolymer synthesized by the "Yamazaki" Route that was heated upto 425°C at a heating rate of 10°C/minute.

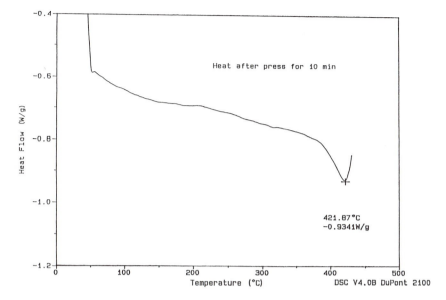

Figure 3. The Differential Scanning Calorimetry thermogram at a heating rate of 10°C/minutes of a representative poly(amide-imide) homopolymer synthesized by the "Yamazaki" Route that was annealed in the hot press at 325°C for 10 minutes.

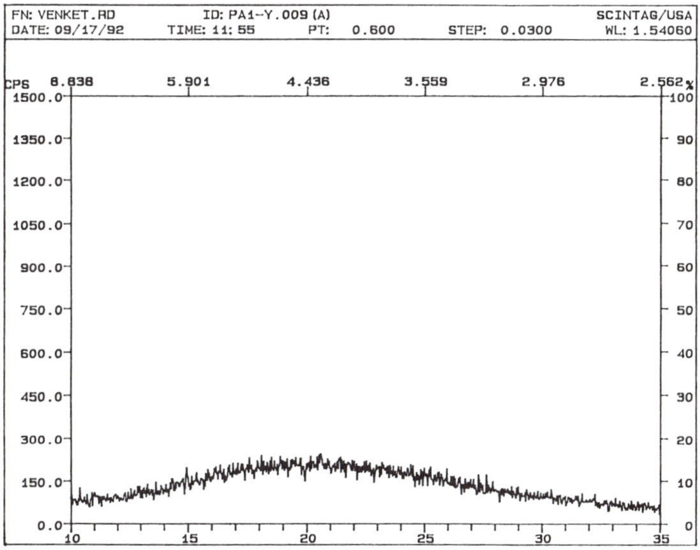

Figure 4. The Wide Angle X-Ray Diffractogram of an "as made" poly(amide-imide) homopolymer.

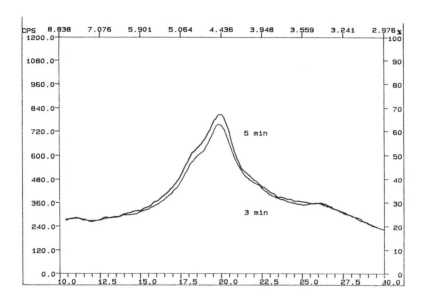

Figure 5. The influence of annealing a poly(amide-imide) homopolymer on the Wide Angle X-Ray Diffraction pattern.

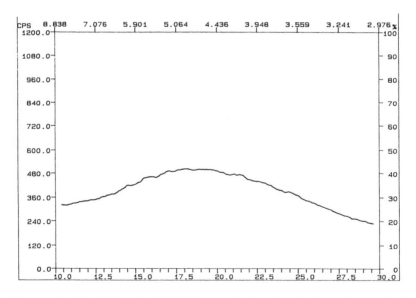

Figure 6. The Wide Angle X-Ray analysis of melted and quenched poly(amide-imide) homopolymer.

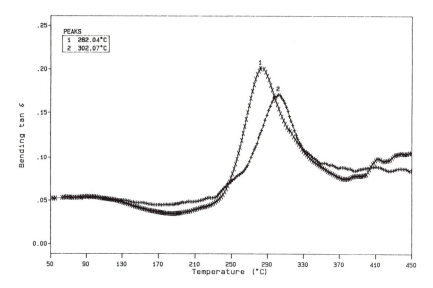

Figure 7. The influence of annealing on the tan δ of a poly(amide-imide) homopolymer.

provides insight into the complex effects occurring at high temperatures. After about 400°C, the storage modulus shows a rise in value subsequent to a marginal drop due to the melting process. This rise in the modulus may be attributed to cross-linking that could occur at these temperatures. Samples quenched from above the melting temperature have apparently undergone enough cross-linking to prevent recrystallization of the system at typical annealing temperatures. This underscores the limitations of melt processing of these semi-crystalline poly(amide-imides) with T_m values greater than 400°C.

In order to develop, a melt-processible system, *4,4' ODA* was replaced in part by *1,3-* phenylene diamine [**mPDA**] to the extent of 50 mole %. The function of **mPDA** would be to break up the all-para sequences and to introduce meta "kinks" in the chain thereby reducing both, chain rigidity as well as order. The synthesis of such melt-processable polymers is described in **Scheme 2.**

The resulting copolymers were now melt-processible. Tough, compression molded plaques were prepared for DMA analysis using the following procedure: The platens were first heated to a temperature of 325°C. The polymer sample was then charges into the mold and subjected to a nominal pressure of about 0.5 ksi for 1 minute. Upon verification of the onset of flow of the material within the mold, the platens were closed again. The material was then compressed with a pressure of 10 ksi for 10 minutes, opening the mold frequently to facilitate the escape of any volatiles thereby ensuring the absence of voids in the molded sample. The temperature was maintained at 325°C throughout the entire molding operation. Clear, transparent, tough plaques were obtained using the molding cycle described above.

Interestingly, the T_g values did not drop to an appreciable extent due to incorporation of *1,3-* phenylene diamine, as is clear from the DMA scan shown in **Figure 8**. The temperature window for melt processing these materials was

$[\eta]$ = 0.7 dl/g. (NMP, 25°C)

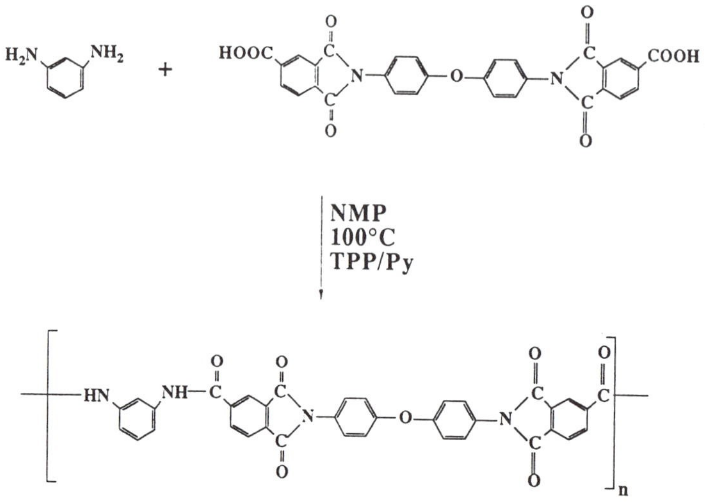

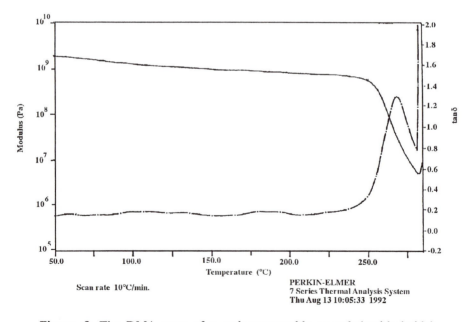

Scheme 2. Synthesis of melt-processible poly(amide-imides) using the Yamazaki Route.

Figure 8. The DMA scan of a melt-processable co-poly(amide-imide) synthesized by the "Yamazaki" Route at a scan rate of 10°C / minute and at a frequency of 1.0 Hz.

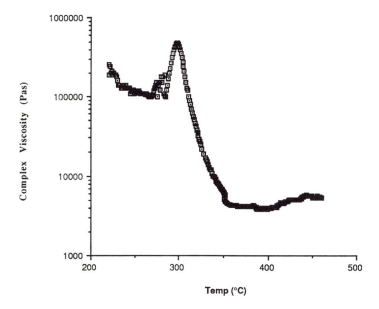

Figure 9. The variation in the complex melt viscosity of a melt-processable poly(amide-imide) with temperature

demonstrated by the dynamic scan of the complex melt viscosity as a function of temperature using a Bohlin parallel plate type rheometer. The scan was run in air at a frequency of 1.5 Hz at a heating rate of 3°C/min. This is as shown in **Figure 9**. It can be seen that the melt viscosity commences to drop above 300°C and attains a minimum value at about 350°C. A stable, low viscosity is maintained till a temperature of about 400°C. Above 400°C, the viscosity begins to rise again as a result of chemo-rheological crosslinking of the polymer. These materials can therefore be melt-processed in the temperature range 350~400°C. Thus, once the tendency of poly(amide-imides) to crystallize has been eliminated, melt-processable systems could be generated.

Conclusions

This investigation has afforded a better understanding of the complex effects at typical molding temperatures of poly(amide-imides). Thus, it may be stated that crystallization and chemorheological phenomena account for the poor melt flow and processibility of the **ODA**-based poly(amide-imides) homopolymers. Absence of order in the "as-made" materials does not prevent the development of crystallinity at typical molding temperatures. The synthesis of appropriate amorphous copolymers of controlled molecular weight and end groups has been optimized and these could be melt-fabricated.

Acknowledgements

The authors appreciate the support of this research by NSF Science & Technology Center at Virginia Tech under Contract DMR-912004. They also thank the Amoco Chemical company for early support of the project.

References

(1)   W.M.Alvino & L.W.Frost, *J. Polym. Sc.*, **Part A-1**, **9**, 2209, 1971; D.C.Clagett, *Encyclopedia Poly. Sci. Eng.*, **6**, 104, John Wiley & Sons, N.Y., 1985.

(2)   K.L.Mittal, Ed., *"Polyimides: Synthesis, Characterization & Applications"*, Vols. 1&2, Plenum, 1984.

(3)   C.Feger, M.M.Khojasteh & J.E.McGrath, Eds., *"Polyimides: Materials, Chemistry & Characterization"* Elsevier Science Publishers B.V., Amsterdam, 1989; P.M.Hergenrother, H.Stenzenberger & D.H.Wilson, *"Polyimides"* Chapman Hall, N.Y., (1990).

(4)   P.R.Young & A.C.Chang, *SAMPE J.*, **22**, 70, 1986.

(5)   E.Sacher, *J.Macromol. Sci. Phys.*, **B25**, 405, 1986.

(6)   R.W.Snyder, B.Thomson B.Bartges, D.Czerniawski & P.C.Painter, *Macromolecules*, **22**, 4166, 1989.

(7)   S.I.Kuroda & I.Mita, *Eur.Polym.J.*, **25**, 611, 1989.

(8)   R.O.Waldbauer, M.E.Rogers, C.A.Arnold, G.A.York, Y.J.Kim & J.E.McGrath, *Polym. Prepr.*, **31(2)**, 432, 1990.

(9)   M.E.Rogers, H.Woodard, A.Brennan, P.M.Cham, H.Marand and J.E.McGrath, *ACS Polym. Prepr.*, **33(1)**, 461, 1992; M.E.Rogers, M.H.Woodard, J.E.McGrath, & A.Brennan, *Polymer*, **34**, 849, 1993.

(10) J.E.McGrath, M.E.Rogers, C.A.Arnold, Y.J.Kim & J.C.Hedrick, *Makromol. Chem., Macromol. Symp.*, **51**, 103, 1991.

(11) Y.J.Kim, T.E.Glass, G.D.Lyle, & J.E.McGrath, *Macromolecules*, **26**, 1344, 1993.

(12) N.Yamazaki, F.Higashi and J.Kawabata, *J. Polym. Sci., Polym. Chem. Ed.*, **12**, 2149, 1974.

(13) F.Higashi, Y.Taguchi, N.Kokubo and H.Ohta, *J. Polym. Sci., Polym. Chem. Ed.*, **19**, 2745, 1981.

(14) N.Yamazaki, M.Matsumoto and F.Higashi, *J. Polym. Sci., Polym. Chem. Ed.*, **13**, 1373, 1975.

(15) C.P.Yang and S.H.Hsiao, *Makromol. Chem.*, **190**, 2119, 1989.

(16) C.P.Yang and S.H.Hsiao, *J. Polym. Sci., Polym. Chem. Ed.*, **28**, 1149, 1990.

(17) C.P.Yang and S.H.Hsiao, *Makromol. Chem.*, **191**, 155, 1990.

(18) S.H.Hsiao and C.P.Yang, *J. Polym. Sci., Polym. Chem. Ed.*, **29**, 447, 1991.

(19) C.P.Yang, J.H.Lin and S.H.Hsiao, *J. Polym. Sci., Polym. Chem. Ed.*, **29**, 1182, 1991.

(20) W.Wrasidlo and J.M.Augl, *J. Polym. Sci., Polym. Chem. Ed.*, **7**, 321, 1969.

(21) J.D.Summers, C.A.Arnold, R.H.Bott, L.T.Taylor, T.C.Ward and J.E.McGrath, *SAMPE Intl. Symp.*, **32**, 613, 1987.

(22) J.D.Summers, Ph.D. Thesis, Virginia Polytechnic Institute and State University, Blacksburg, VA, 1988.

(23) C.A.Arnold, J.D.Summers, Y.P.Chen, R.H.Bott, D.Chen and J.E.McGrath, *Polymer*, **30**, 986, 1989.

(24) D.L.Wilkens, C.A.Arnold, M.J.Jurek, M.E.Rogers and J.E.McGrath, *J. Thermoplas. Comp. Mater.*, **3(1)**, 4, 1990.

(25) V.N.Sekharipuram, M.A.Vrana, S.S.Joardar, M.Konas, A.R.Shultz, T.C.Ward & J.E.McGrath, *ACS Polym. Prepr.*, **34(1)**, 620, 1993; Manuscript in preparation to be submitted to *J. Polym. Sci. Polym. Chem* .

RECEIVED November 11, 1994

Chapter 15

Metal Salt–Polymer Composites

Complexation of Metal Salts with the Phosphorus–Oxygen Bond in Poly(arylene ether phosphine oxide)s

E. Bonaplata[1], C. D. Smith[2], and J. E. McGrath[3]

Department of Chemistry and National Science Foundation Science and Technology Center, High Performance Polymeric Adhesives and Composites, Virginia Polytechnic Institute and State University, Blacksburg, VA 24061–0344

Controlled high molecular weight poly(arylene ether phosphine oxide)s were prepared via the nucleophilic aromatic substitution route. These materials were subsequently used in the preparation of polymer/metal composite films. It was demonstrated that a variety of metal salts including metals such as iron, zinc, cobalt, and copper, can be complexed at a molecular level with the phosphoryl group in films of these polymers producing novel transparent metal halide/polymer "composites". A procedure for obtaining homogeneous films from solutions of the metal halide complexed polymers has been developed. The effect of chemical composition of the chain, type of metal salt, molar concentration of the metal salt, and heating cycle were investigated and found to influence properties of the films such as solubility, glass transition temperature, thermal stability, and storage modulus. In addition, FTIR experiments as well as T_1 phosphorus (^{31}P) NMR measurements were conducted to demonstrate the existence of metal complexation in the solid state.

Poly(arylene ether phosphine oxide)s (PEPO) are a recently identified sub-set of an important macromolecular series which includes industrially important high performance thermoplastics, such as the polysulfones, e.g. UDEL® and polyether ketones, e.g. PEEK®, PEKK®, etc. The PEPO materials show elevated glass transition temperature, high thermal and oxidative stability, improved solubility, and increased flame resistance.[1-4] The present work has focused on the study of the complexation of several metal salts with the phosphoryl group in poly(arylene ether phosphine oxide)s. The ultimate goal is the preparation of easily processed polymer/metal salt composites with novel properties (electrical, thermal, surface resistivity, permeability, etc.), while maintaining the optimal characteristics of the base materials.

The formation of stable complexes of low molecular weight organophosphine oxides with a large variety of metals was reviewed by Karayannis and coworkers.[5]

[1]Current address: Dr. Bastos 42, 2 B, Majadahonda, 28220 Madrid, Spain
[2]Current address: Air Products & Chemicals, Inc., 7201 Hamilton Boulevard, Allentown, PA 18105
[3]Corresponding author

0097–6156/95/0603–0227$12.00/0

Preliminary studies conducted in our laboratories demonstrated the potential for complexation of poly(arylene ether phosphine oxide)s with a number of metal salts.[6] A series of composite films were prepared in order to further investigate the metal coordination capabilities of these phosphine oxide containing macromolecules. The films were prepared from homogeneous DMAc solutions of various metal salts (FeCl$_3$, CoCl$_2$, and CuCl$_2$).

Comparative studies were conducted on 20 Kg/mol PEPO films containing varying amounts of the different metal chlorides. Bis A-PEPO films comprising 0-20 mole % (based on the repeat unit molecular weight, 502.55 g/mol) of FeCl$_3$, CoCl$_2$, or CuCl$_2$ were prepared in order to investigate the variation in properties such as solubility, thermal stability, and glass transition temperature with the type and content of metal salt. Additionally, BP-PEPO composite films with an equal loading of the metal halides (20 mole %) were prepared to study the influence of polymer backbone chemistry.

EXPERIMENTAL

Solvents and General Reagents: The dipolar aprotic solvent (N,N'-dimethylacetamide) (DMAC) used in the polymerization reaction as well as in the composite film preparation was vacuum distilled over calcium hydride and stored in an anhydrous environment prior to use. Potassium carbonate was obtained from Fisher and used as supplied. Cobalt(II)chloride, copper(II)chloride, iron(III)chloride, and zinc(II)chloride were obtained from Aldrich and employed without further purification. Toluene (Fisher) and methanol (Fisher) were used as received.

Monomers: Monomer grade bisphenol A (Bis A) was kindly supplied by Dow Chemical and required no further purification.   Biphenol (Aldrich) was recrystallized from saturated deoxygenated acetone, dried over a nitrogen flow, crushed and vacuum dried at 50°C. Bis(4-fluorophenyl)phenyl phosphine oxide) (BFPPO) was prepared and purified by known Grignard techniques.[6]

Synthesis of High Molecular Weight Poly(Arylene Ether Phosphine Oxide)s: In a typical procedure for the preparation of high molecular weight poly(arylene ether phosphine oxide)s a 250 ml 4-necked round bottom flask equipped with a nitrogen inlet, an overhead stirrer, a Dean-Stark trap with condenser and a thermometer was charged with BFPPO and a slight excess of one of two bisphenols, bis-A or biphenol, in order to obtain phenolic endgroups and control the molecular weight. For example, 9.085 g (0.040 mol) of bis-A were weighed and combined with 12.197 g (0.039 mol) BFPPO. The teflon beakers from which the monomers were transferred were rinsed well into the round bottom flask with DMAc, for a total volume of 100 ml (20% (w/w) solids). A 15% excess of K$_2$CO$_3$ (6.17 g, 0.045 mol) and 45 ml toluene were combined with the reaction mixture. The reaction temperature was controlled by a high temperature silicone oil bath, and a constant purge of nitrogen was maintained. After the water and toluene azeotrope formed at about 145°C, the system was allowed to dehydrate for four hours. Next, the temperature of the mixture was raised to 155-160°C and held there for a minimum of twelve hours. The resulting viscous solution was dark brown with a suspension of the white inorganic salt. It was allowed to cool, diluted with chloroform, and filtered through a Buchner funnel to remove the inorganic salts. The solution was neutralized using glacial acetic acid to afford a viscous clear brown mixture. This solution was precipitated into a 80:20 methanol:water mixture using a high speed blender. The highly fibrous white material was dried in a vacuum oven at 100°C for approximately 12 hours, dissolved in chloroform, filtered, neutralized, precipitated in pure methanol and dried a second time under the same conditions. The resulting polymers were soluble in chloroform, methylene chloride, tetrahydrofuran, and polar aprotic solvents (e.g., DMAC, NMP).

Preparation of Metal Halide-Poly(Arylene Ether Phosphine Oxide) Complexed Films: The amount of materials for a given composition of metal to polymer ratio were calculated employing the equation:

Mass of polymer used $*$ Mole % metal $*$ FW metal halide = g of metal halide to be used   MW PEPO repeat unit   100

For example, for 20 mole% $CoCl_2$ in BP-PEPO, 1.0 g of 20,000 g/mol hydroxy endcapped BP-PEPO (molecular weight of repeat unit = 460.28 g/mol), 0.0564 g $CoCl_2$ (F. W. = 129.84 g/mol), and 7 ml of DMAc were used. Once calculated, the required amounts were weighed into a vial and dissolved in DMAc (~ 15 % (w/w) solids). The solution was stirred for several hours or until homogeneous, and filtered to remove any dirt particles. Under a low flow of N_2, the solutions were poured onto a silanated glass plate and placed in a closed glass box.

The solvent was slowly evaporated at 65°C using an infrared heat lamp until the film began to pull off the glass plate (12 hours). The film was subsequently placed in a vacuum oven at 25°C for at least two hours, after which the temperature was very slowly raised to just below the glass transition temperature of the unmodified poly(arylene ether phosphine oxide), and maintained there for several hours in order to ensure the removal of any trapped solvent.

Characterization Methods: Intrinsic viscosities were measured at 25°C in a Cannon-Ubbelohde viscometer, generally using chloroform as a polymer solvent. Fourier transform infrared spectra (FTIR) were obtained with a Nicholet MX-1 instrument. The polymer spectra were taken as very thin films on an FTIR film holder.   Solution (^{31}P) NMR spectra were obtained on a Varian 400 MHz instrument and solid state (^{31}P) NMR spectra were obtained on a Bruker MSL 300 instrument..   All spectra were referenced to 85% H_3PO_4 at 0 ppm, and dichlorophenyl phosphine sulfide was used as a standard (76.05 ppm). Thermogravimetric analysis (TGA) and differential scanning calorimetry (DSC)were conducted on the Perkin-Elmer 7 series thermal analysis instrument at a heating rate of 10°C min^{-1}. In the determination of T_g values the samples were heated to 300°C at a heating rate of 10°C min^{-1}, rapidly cooled to 100°C, and subsequently heated to 300°C at 10°C min^{-1}. The glass transition temperatures (T_g) were recorded as the midpoint of the second heat. Dynamic mechanical analysis (DMA) was conducted on the Polymer Lab. Mark II DMTA in the 3 point bending mode at a frequency of 1 Hz, and at a heating rate of 2.5°C per minute.

Results and Discussion

Two high molecular weight poly(arylene ether phosphine oxide)s (Bis A-PEPO and BP-PEPO) were synthesized via nucleophilic aromatic substitution.[7-9]   These polymers have very attractive properties such as good thermo-oxidative stability, hydrolytic stability and flame resistance, as well as the potential for metal complexation through the phosphine oxide moiety.[1-4,6] The structures are provided in Scheme 1.

In general, the films prepared from solution were homogeneous, clear, and tough with colors varying from brown in the case of the iron and yellow for the copper to deep blue for the cobalt. The films containing from 0 to 15 mole % of the metal salts all redissolved in DMAc, N-methylpyrrolidone, and chloroform at ambient temperature. However, when 20 mole % of any of the three salts was incorporated the room temperature solubility was limited to amide solvents.

The effect of metal salt incorporation on the thermal stability in air and in nitrogen as well as the glass transition temperature of the poly(arylene ether phosphine oxide)s were investigated. The thermo-oxidative stability in terms of the 5 % weight loss temperature of the $FeCl_3$ films decreases relative to the unmodified Bis A-PEPO polymer with increasing metal content. In the case of the $CoCl_2$ complexes the 5 and 10% films show a higher 5 % weight loss temperature relative to the unmodified Bis A-PEPO. However, the 5 % weight loss of the 15 and 20 mole % films were evidenced at lower temperatures in comparison to the base polymer. It is suspected that the latter behavior could possibly be influenced by a small amount of residual solvent. The copper modified films show the greatest lowering in thermal stability relative to the base polymer. The dynamic TGA thermograms at 10°C/min in air of the $FeCl_3$/Bis A-PEPO composite films are shown in Figure 1 as an example.

Similar trends in thermal stabilities were observed in nitrogen. It needs to be noted that Fe(III) chloride and Cu(II) chloride decompose above 300°C, whereas, Co(II) chloride is stable to over 500°C (dynamic TGA). Hence, the observed lowering of the thermal stabilities in the composite films may result from a combination of partial decomposition of the salts, and the influence of metal complexation.

The char yields at 750°C of the complexes are in all cases higher than that of the unmodified polymer. The enhanced char yields in the metal modified films, in almost all cases, exceed the amounts of metal salt initially present in the films. Thus, the increase in the char yields cannot be attributed solely to the weight of the metal salt.

The thermo-oxidative stability of the 20 mole % metal chloride/BP-PEPO composite films was also found to depend on the chemistry of the metal salt. The 5 % weight loss temperature of unmodified BP-PEPO was increased by the incorporation of either $FeCl_3$ or $CoCl_2$, and decreased by the addition of $CuCl_2$. In contrast, in the case of Bis A-PEPO the 5 % weight loss temperature was increased by the incorporation of $CoCl_2$, and decreased by the addition of $FeCl_3$ or $CuCl_2$. In general, the thermal stabilities of the metal modified polymers are influenced by the backbone structure of the polymer, the chemistry of the metal salt, and their relative concentrations.

The glass transition temperatures of the metal complexed films were also found to depend on these three factors. The glass transition temperatures were measured by differential scanning calorimetry (DSC) and by dynamic mechanical analysis (Discussed in the next section). The glass transition temperatures of the Fe(III) chloride/Bis A-PEPO composite films gradually increase as a function of metal halide concentration. An increase in the T_g was also observed for the 5 and 10 mole % Co(II) chloride/Bis A-PEPO films. However, upon loading of 15 and 20 mole % of this salt, the T_g is depressed relative to the base material. All the Cu(II) chloride/Bis A-PEPO films exhibit a lowering of the glass transition temperature in comparison to the unmodified material.

An increase in the DSC glass transition temperature such as that evidenced for $FeCl_3$/Bis A-PEPO composites, strongly suggests the existence of strong interactions between the metal salt and the polymer which may decrease the flexibility of the chains. The complexation with $CuCl_2$, on the other hand, surprisingly appears to depress the T_g of the Bis A-PEPO film. However, interchain complexation is suggested by the much higher rubbery state modulus as can be seen in the dynamic mechanical spectra (Figure 3). The observed depression of the glass transition temperature could possibly result from the combination of a

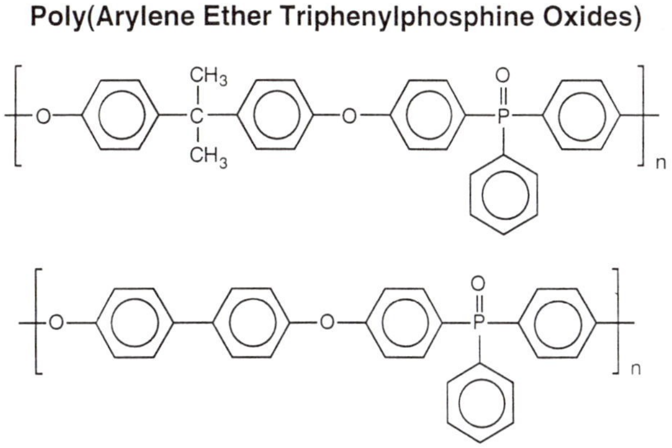

Poly(Arylene Ether Triphenylphosphine Oxides)

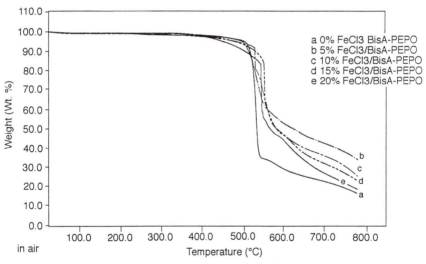

Scheme 1

Figure 1. Dynamic TGA Thermograms in Air of FeCl₃/Bis A-PEPO Composite Films (10°C/min)

Figure 1. Dynamic TGA Thermograms in Air of FeCl$_3$/Bis A-PEPO Composite Films (10°C/min)

decrease in the intermolecular forces as a result of complexation and a plasticizer effect by uncomplexed metal salt. The complexation behavior of the $CoCl_2$ salt is suggested to be similar to that of the $FeCl_3$, as judged by NMR. The lower T_gs observed for the higher $CoCl_2$ concentration films may result from the presence of residual solvent, although this could not be confirmed by TGA. This aspect was discussed in relation to the thermal stability of the composites and is in agreement with the trends observed for the BP-PEPO films discussed below.

The glass transition temperatures of the 20 mole % metal chloride/BP-PEPO composites in relation to that of the unmodified film are presented in Figure 2. An increase in the T_g is observed for the $FeCl_3$ and $CoCl_2$ modified films, whereas, a depression is experienced as a result of $CuCl_2$ modification. These trends in T_gs are influenced by factors including crosslinking through complexation, enhanced chain rigidity, decrease of preexisting intermolecular forces and plasticizer effects.

The mechanical properties of the metal salt-BP-PEPO complexes were investigated using dynamic mechanical analysis (DMA) in the three point bending mode. The behavior of the 20 mole % metal chloride BP-PEPO films along with that of the unmodified polymer film are shown in Figure 3. The metal salt modified films show substantially improved modulus relative to the base polymer. The storage modulus of the unmodified BP-PEPO film decreases gradually with increasing temperature and loses its mechanical integrity at T_g or slightly above. On the other hand, the metal salt systems exhibit a retention of the storage modulus well beyond the glass transition temperature. In fact, the storage modulus of the $CoCl_2$ film remains essentially unchanged through the glass transition.

This behavior is highly characteristic of crosslinked networks and strongly suggests the existence of interchain, perhaps ionomeric, connectivity through the metal. It is important to emphasize again that this network formation is chemically reversible since the metal-polymer composites redissolve in amide solvents. The softening temperatures of the polymeric complexes obtained from the tan δ values (Figure 3) follow the same trends observed by differential scanning calorimetry.

In order to substantiate the presence of interactions between the metal salt and the polymer matrix through the phosphine oxide linkage, thin films of the BP-PEPO modified, base materials were cast and their FTIR spectra obtained. The phosphine oxide bond stretch appears at approximately 1197 cm^{-1}. The intensity of the phosphine oxide stretch was found to decrease with the incorporation of 20 mole % of the metal halides relative to the unmodified BP-PEPO film. The observed intensity decrease of the phosphine oxide peak can possibly result from the complexation of a fraction of the phosphine oxide moieties in the chains. The appearance of a new peak at lower wavenumbers corresponding to the complexed moieties was not readily apparent from the spectra. It is possible that this low intensity peak is hidden beneath other peaks. Therefore, it was desirable to verify the existence of the complexation by other methods.

Phosphorus (^{31}P) studies were undertaken to further substantiate the interaction of the phosphine oxide group in the polymer with the metal salts. Solutions of Bis A-PEPO in DMAc containing varying mole percents (0-20 mole % based on the repeat unit molecular weight, 502.55 g/mol) of $FeCl_3$ (Fe^{III}, d^5), $CoCl_2$, (Co^{II}, d^7) and $CuCl_2$ (Cu^{II}, d^9) were prepared. A marked downfield shift and broadening of the phosphine oxide peak was observed on binding with the metal salts. This effect is magnified with increasing metal salt content as is seen for Bis A-PEPO solutions with different $CuCl_2$ concentrations in Figure 4. The same trend is observed for $FeCl_3$ and $CoCl_2$. The greatest shifts were observed for the $CoCl_2$/Bis A-PEPO solutions with those of the $CuCl_2$ and $FeCl_3$ being smaller and of similar magnitude.

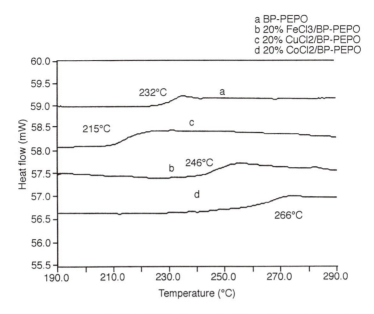

Figure 2. DSC Traces for BP-PEPO Composite Films (Second Scan. 10°C/min) (% = mole % based on repeat unit molecular weight)

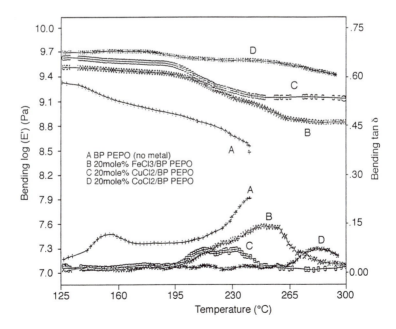

Figure 3. Dynamic Mechanical Analysis of BP-PEPO Composites (1Hz., 2.5°C/min)

The metal halides investigated are all paramagnetic. Iron(III) is a d^5 transition metal, cobalt(II) is d^7, and copper(II) is d^9. In all three complexes of these metal chlorides with triphenyl phosphine oxide the transition metal is found to be in the tetrahedral high spin state. Here, the iron has five unpaired electrons, the cobalt three, and the copper one. The presence of unpaired electrons in the polymeric complexes was verified by electron paramagnetic resonance (EPR). The shift observed in solution due to the interaction of a paramagnetic substance with a diamagnetic reagent is a well studied phenomenon and is discussed in many texts. It results from a partially covalent bond or weak transient complex formation that causes a finite unpaired electron spin density on the nucleus. Alternatively, ligand exchange in solution could account for the broad NMR linewidths observed. Two resonances would be predicted from the complexed and free P=O units in the polymer chain. However, if the exchange of the metal between different phosphine oxides in the solution occurs at a rate comparable with the NMR time scale (10^{-1}-10^{-6} sec.) only one broad resonance is observed.[10] This seems to be the case in our systems.

Solvent interference with the metal salt-polymer coordination undoubtedly occurs in the DMAc solutions. Therefore, the solid state ^{31}P NMR spectra of the composite films were investigated. The paramagnetic shift and peak broadening were not observed in the solid state probably due to the restriction of motion in this state. However, ^{31}P NMR spin-lattice relaxation time (T_1) measurements at room temperature have proved useful in the study of the direct interaction of the paramagnetic salts with the phosphine oxide polymers. The relaxation times for the BP-PEPO films were determined using the inversion recovery method.[10] The room temperature T_1 values of the phosphine oxide resonance were measured for the $FeCl_3$, $CoCl_2$, and $CuCl_2$ modified BP-PEPO paramagnetic composite films as well as for a 20 mole % $ZnCl_2$/BP-PEPO composite film which was used as a control. The T_1s of the paramagnetic composites were significantly lowered relative to the unmodified BP-PEPO film with the largest decrease being observed for the $FeCl_3$ composite followed by that of the $CoCl_2$ and, lastly, the $CuCl_2$ film. On the other hand, within experimental error, the T_1 of the diamagnetic system was the same as for the BP-PEPO film.

Paramagnetic metal ions generally enhance the NMR relaxation rates, therefore, reducing the spin relaxation times. In the vicinity of unpaired electrons the relaxation process is dominated by the electron-nuclear dipole-dipole interactions.[11] The relaxation rates ($1/T_1$) for the direct interaction of a nuclear spin with a paramagnetic ion are related to the number of unpaired electrons as well as the separation between the electron spin and the nuclear spin (r).[11] The relaxation rate is directly proportional to the total electron spin [$S(S+1)$, where S is the number of unpaired electrons], and increases as r^{-6}. Assuming that the coordination geometry of the polymeric complexes is the same as that of the monomeric counterparts, the number of unpaired electrons for Fe(III), Co(II), and Cu(II) are five, three, and one, respectively. The trend observed in the reduction of the relaxation times for the paramagnetic salts appears to be consistent with the dependence of $1/T_1$ on the total electron spin. Additionally, this finding substantiates that the direct interaction between the metal chloride and the phosphine oxide unit in the polymer is responsible for the observed chemical shifts in solution.

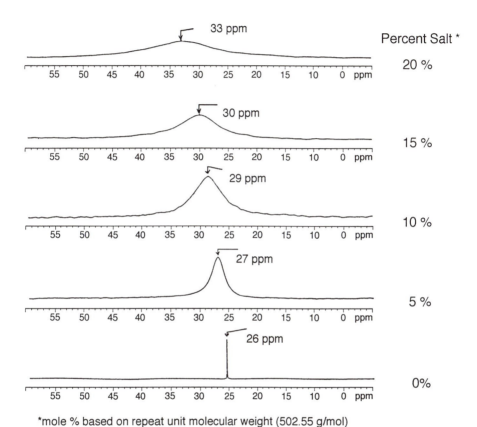

*mole % based on repeat unit molecular weight (502.55 g/mol)

Figure 4. Phosphorus (31P) NMR of CuCl$_2$/Bis A-PEPO in DMAc Referenced to H$_3$PO$_4$ at 0 ppm (161.9 MHz)

Conclusions

Homogeneous, clear, tough PEPO/metal salt films were obtained with their colors varying from brown in the case of the iron and yellow for the copper to deep blue for the cobalt. The unmodified films were soluble in DMAc, N-methylpyrrolidone, and chloroform; however, the solubility of the complexed films was limited to amide solvents. Each metal chloride/polymer pair appears to be a case unto itself regarding specific polymer properties such as thermal stability and glass transition temperature.

TGA scans revealed that the temperature related to the onset of decomposition of unmodified Bis A-PEPO was increased by the incorporation of $CoCl_2$, and decreased by the addition of $FeCl_3$ or $CuCl_2$. In the cases for which a lowering in thermal stability was observed, the decrease was generally more pronounced as the amount of metal salt added was increased. In contrast, in the case of BP-PEPO the 5 % weight loss temperature was increased by the incorporation 20 mole % of either $FeCl_3$ or $CoCl_2$, and decreased by the addition of $CuCl_2$. In general, the thermal stabilities of the metal modified polymers are influenced by the backbone structure of the polymer, the chemistry of the metal salt, and their relative concentrations. The glass transition temperatures of the $FeCl_3$ and of the $CoCl_2$ polymer films gradually increase with increasing salt content in comparison to the unmodified polymers. Whereas, all $CuCl_2$ films exhibit a depression of the glass transition temperature relative to the base material. These trends in T_gs are influenced by factors including crosslinking through complexation, enhanced chain rigidity, decrease of preexisting intermolecular forces, and plasticizer effects.

The dynamic mechanical analysis of the biphenol PEPO films with 20 mole % concentration of $CoCl_2$, $FeCl_3$ and $CuCl_2$, respectively, showed a remarkable increase and retention of the storage modulus as a result of the incorporation of metal salt. This rise in the storage modulus may be due to ionomer-like cross-linking by the metal, but is possibly related to the solid state composite-like structure of the new hybrid. Moreover, FTIR and Phosphorus (^{31}P) NMR studies of the composites of all three metal chlorides are consistent with the existence of direct complexation between the phosphine oxide moiety in the polymer backbone and the metal halide, in solution as well as in the solid state. The environmental stability of the complexes has not been studied, but cobalt chloride-PEPO films show no major changes after more than six months under ambient conditions.

Acknowledgments

The authors would like to thank the Virginia Institute for Materials Systems, the Office of Naval Research (ONR - Contract No. N00012-91-5-1037) and the NSF Science and Technology Center on High Performance Polymeric Adhesives and Composites for support (DMR 91-20004). Generous support form the GenCorp Foundation is also appreciated.

Literature Cited

1.   Clagett, D. C. in *Encyclopedia of Polymer Science and Engineering*; H. F. Mark, N. M. Bikales, C. G. Overberger and G. Menges, Ed.; John Wiley and Sons: New York, 1986; Vol. 6; pp 94.
2.   Johnson, R. N. in *Encyclopedia of Polymer Science and Technology*; N. M. Bikales, Ed.; John Wiley and Sons: New York, 1969.
3.   Johnson, R. N.; Farnham, A. G., *J. Polym. Sci., Polym. Chem. Ed.* **5**, 2415 (1967).

4. Searly, O. B.; Pfeiffer, R. H., *Polym. Eng. Sci.* **25** , 474 (1985).
5. Karayannis, N. M.; Mikulski, C. M.; Pytlewski, L. L., *Inorg. Chi. Acta, Rev.,* 69 (1971).
6. Smith, C. D., Ph. D. Thesis, Virginia Polytechnic Institute and State University, 1991.
7. Robeson, L. M.; Farnham, A. G.; and McGrath, J. E., *Appl. Polym. Symp.,* 1975, **26**. 373.
8. Viswanathan, R. and McGrath, J. E., *Polymer Preprints,* 20(2) 365, 1979; Viswanathan, R., Ph.D. Thesis, Virginia Polytechnic Institute and State University, 1981; Viswanathan, R.; Johnson, B. C.; and McGrath, J. E., *Polymer,* 25(12) 1827 (1984).
9. Blinne, G.; Cordes, C., *Chem. Abstr.* **90**, 138421 (1979) (*Ger. Pat.* 2 731 816 (1979)).
10. Crabtree, R. H. in *The Organometallic Chemistry of the Transition Metals*; John Wiley & Sons: New York, 1988, pp 225.
11. Krugh, T. R. in *Molecular Biology. An International Series of Monographs and Textbooks*; L. J. Berliner, Ed.; Academic Press: New York, 1976; pp 339.

RECEIVED February 17, 1995

Chapter 16

Refractive Indices of Aromatic Polyimides

David C. Rich[1], Peggy Cebe[1], and Anne K. St. Clair[2]

[1]Department of Materials Science and Engineering, Massachusetts Institute of Technology, Cambridge, MA 02139
[2]Materials Division, NASA Langley Research Center, Hampton, VA 23681-0001

Experimentally measured refractive indices of polyimides of varying chemistry are compared to predicted values calculated from the methods of van Krevelen (1) and Bicerano (2). The refractive index reducing effects of fluorinated substituents, such as the 6F and SF_5 groups, are of particular interest. The prediction techniques successfully estimate the decrease in refractive index with increasing fluorine content that has been widely observed experimentally. In addition, while measured refractive indices of non-fluorinated polyimides generally deviate above predicted values, measured refractive indices of polyimides with high fluorine content are very close to predicted values. Disparities between measured and predicted refractive indices are explained in terms of crystallinity and inter-chain interaction, both of which are expected to be more prevalent in the non-fluorinated materials.

Aromatic polyimides are used in electronic packaging, aerospace applications, and other high technologies requiring high temperature stability, solvent resistance, and excellent mechanical and electrical properties. Recently, novel polyimides have been synthesized which have low refractive indices. The materials are optically transparent and suitable as coatings and waveguides. One strategy for achieving low refractive index has been incorporation of fluorine-containing groups. Fluorinated polyimides exhibit low dielectric constants and refractive indices. It is unclear, however, exactly what mechanisms cause this reduction. This work compares the refractive indices of non-fluorinated and fluorinated polyimides.

We use two property prediction methods to estimate the expected effect of chemical modification on refractive index. It is assumed that these estimations account for only the intrachain electronic polarizability of the polymer molecule. Predicted properties are then compared to experimentally measured refractive indices.

Property Prediction Methods

Van Krevelen/Vogel Method. Van Krevelen's method (1) is an additive group contribution technique. The polymer backbone is separated into component functional groups, each of which contributes an additive tabulated value to the overall Vogel molar refraction, R_V. Refractive index, n, is calculated from:

$$R_V = n\,M \tag{1}$$

where M is the molecular weight of the repeat unit in grams/mole. Functional groups in different bonding environments, for example, when attached to a phenyl group as opposed to an alkyl group, often have different tabulated values. Correction factors are also tabulated for ring structures, and for aromatic substitution when there are more than two substitutions. Since a correction for imide rings is not tabulated, a value for similar rings is used.

Bicerano Method. New topological methods for polymer property prediction have become available (2). These methods are applicable to novel chemistries, even those with groups not tabulated by van Krevelen's analysis. The techniques utilize four connectivity indices, which are calculated from the atoms and bonds in the repeat unit. A correlation for refractive index at room temperature has been developed using connectivity indices and several correction numbers defined by Bicerano. Using 183 polymers with refractive indices ranging from 1.34 to 1.71, a standard deviation of about 1% of the average value in the data set was obtained for Bicerano's prediction correlation. No polyimides were included in the original correlation by Bicerano.

Unlike other property prediction methods such as van Krevelen's methods based on the equations of Lorentz-Lorenz and Gladstone-Dale (1), the method of van Krevelen and Vogel and the method of Bicerano do not require experimental measures of the polymer density. They can therefore be used prior to the synthesis of the polymer. However, since they do not take the variable density of the material into account, these two approaches are not applicable to semicrystalline polymers. They would therefore be expected to be most accurate in the case of amorphous polymers.

Experimental

Polyimides ODPA-3,3'-ODA, ODPA-4-BDAF, 6FDA-3,3'-ODA, 6FDA-4-BDAF, 6FDA-DASP, BFDA-DASP, BPDA-DASP, ODPA-DASP, IPAN-DASP, BTDA-DASP, HQDEA-DASP, BDSDA-DASP, BTDA-1,3BABB, PMDA-ODA, and ODPA-PDA were received as thick free standing films from the NASA Langley Research Center. The chemical compositions of these polymers are presented in Tables I-IV. The polyamic acids were synthesized in N,N-dimethylacetamide, solution cast on glass, and cured in a low humidity environment at 100°C, 200°C, and 300°C for one hour at each temperature. PMDA-ODA was annealed an additional 20 minutes at 400°C. Regulus™ NEW-TPI, whose chemical composition is presented in Table I, was received from the Mitsui Toatsu Chemical Co.

Refractive index measurements were made in a Metricon PC-2000 prism coupler, using polarized light of $\lambda = 633$nm, by critical angle determination. Wide angle x-ray scattering was performed using a Rigaku RU300 with a Cu-Kα source, scanning from 3° to 43° 2θ at 3° per minute, sampling at 0.05° intervals.

Anisotropy is taken into account when determining the refractive index of the polymer. Average refractive index (n_{avg}), which is calculated as the average of the measured indices along three perpendicular directions, is utilized. For an in-plane oriented material which is isotropic in the plane of the film, average index is given as:

$$n_{avg} = (1/3)\,(2n_{TE} + n_{TM}) \tag{2}$$

where n_{TE} is the transverse electric (in-plane) index and n_{TM} is the transverse magnetic (out-of-plane) index.

Other refractive index data were taken from available literature (3-9). Details of sample preparation can be found in these sources.

Table I. Dianhydride (R) and diamine (Z) elements of polyimides in Figure 1.

Polyimide (ref.)	R	Z
PMDA-ODA (see also 3)		
BTDA-ODA/mPDA (4)		80% / 20%
ODPA-4,4'ODA (5)		
ODPA-3,3'ODA		
BTDA-1,3BABB (melted/quenched)		
NEW-TPI		
ODPA-APB (5)		

Table II. Dianhydride (R) and diamine (Z) elements of polyimides in Figure 3.

Polyimide (ref.)	R	Z
BPDA-PDA (6)		
PMDA-BPD (7)		
ODPA-PDA		
BTDA-1,3BABBB		

Table III. Diamine (Z) elements of the 6F polyimides in Figure 4.

R: 6FDA =

ODPA =

Polyimide (ref.)	Z	Polyimide (ref.)	Z	Polyimide (ref.)	Z
ODPA-BDAF		6FDA-PDA (9)		6FDA-TFDB (9)	
6FDA-APB (5)		6FDA-DAT (9)		6FDA-TFPDA (9)	
6FDA-3,3'ODA		6FDA-FPDA (9)		6FDA-6FDA (8)	
6FDA-4,4'ODA (8)		6FDA-2DAT (9)		6FDA-OFB (9)	
6FDA-BDAF		6FDA-TFMPDA (9)		6FDA-2TFMPDA (9)	

Table IV. Measured and predicted refractive indices of SF$_5$ polyimides.

Polyimide	R	$n_{meas.}$	$n_{Bic.}$	n_{vK}
6FDA-DASP		1.541	1.536	1.529
BFDA-DASP		1.576	1.559	1.554
BPDA-DASP		1.640/1.635*	1.592	1.582
ODPA-DASP		1.620	1.588	1.576
IPAN-DASP		1.597	1.583	1.570
BTDA-DASP		1.621	1.588	1.577
HQDEA-DASP		1.624	1.594	1.586
BDSDA-DASP		1.647	1.616	1.602

* First value represents side 1, the second value, side 2.

Results and Discussion

Experimental and predicted refractive indices of a wide variety of polyimides were compared. Three types of behavior were found. First, many amorphous, non-fluorinated polyimides were found to have refractive indices of around 1.68-1.70, deviating slightly, though systematically, above the predicted refractive index. Second, some polyimides were found to have refractive indices that were substantially higher than those predicted. Third, fluorinated polyimides had low refractive indices which generally compared favorably to predicted values.

Amorphous, non-fluorinated polyimides.   A comparison between predicted and measured (3-5) refractive indices in nonfluorinated polyimides is shown in Figure 1. The polyimides have generally been arranged from left to right in order of declining measured refractive index. The chemical structures of the polyimides are shown in Table I.

Although the calculated refractive indices are close to experimental values, we notice a systematic underprediction, especially when using the presumably more accurate method of Bicerano. Refractive indices calculated by van Krevelen's method are excellent, underpredicting by only about 0.02-0.03. Bicerano's method underpredicts refractive index by about 0.05-0.06, which is three to four times the standard deviation in Bicerano's correlation. It is unlikely that such large deviations from experimental values in the Bicerano computations can be attributed solely to statistical error.

Charge transfer complexation and/or molecular ordering might result in deviations of predicted from measured refractive indices. There has been no evidence of substantial crystal lattice ordering in these materials. However, as shown in Figure 2a, WAXS of PMDA-ODA may exhibit slight paracrystalline order. Alternatively, chain packing which might result from intermolecular complexation could be responsible for an increase in density of the amorphous phase and/or an increase in polarizability.

Semicrystalline polyimides.   Since crystalline phases are denser than amorphous phases, a measured refractive index substantially higher than that predicted by property prediction methods is an indication that the polymer may be semicrystalline. Figure 3 compares predicted and measured (6,7) indices of polyimides with unusually high refractive indices. See Table II for the chemical structures.

There has been substantial experimental evidence of crystal lattice ordering in these particular materials. For example, WAXS analysis, shown in Figure 2b, confirms the presence of a crystalline phase in our ODPA-PDA sample. Thus, there is likely a strong structure-property relationship between crystallinity and refractive index in polyimides which is not taken into account in the van Krevelen and Bicerano predictions.

Polyimides containing hexafluoroisopropylidene (6F).   The optical properties of polyimides containing the hexafluoroisopropylidene (6F) group have been the subject of much investigation (8-10). It is believed by these authors that the 6F group, which consists of two trifluoromethyl groups bonded to carbon, sterically inhibits charge transfer complexation in the polyimides.

Figure 4 compares predicted and measured (8-10) refractive indices of polyimides containing the 6F group. For the chemical structures, see Table III. Both property prediction methods estimate refractive index quite accurately. Bicerano's method is accurate when the 6F group appears in both the dianhydride and diamine portions of the repeat unit, but very slightly underpredicts refractive index when the 6F group exists in only one segment. This demonstrates that the lower refractive indices

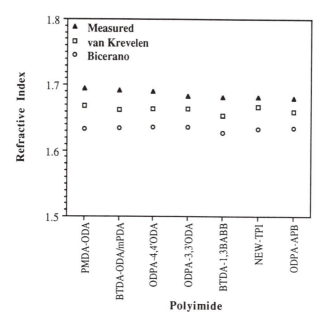

Figure 1. Experimental and predicted refractive indices of non-fluorinated polyimides from Table I.

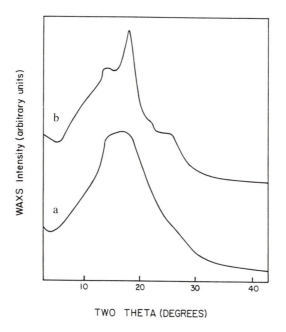

Figure 2. Wide angle x-ray scattering scans: (a) PMDA-ODA, no crystalline reflections, (b) ODPA-PDA, crystalline reflections.

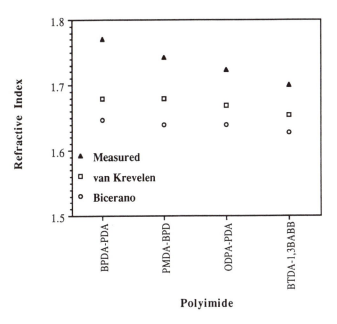

Figure 3. Measured and predicted refractive indices of semicrystalline polyimides.

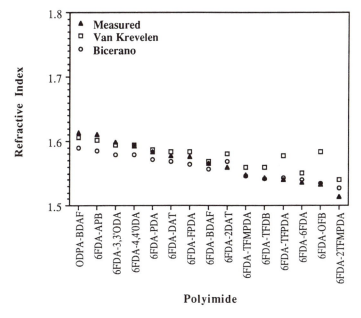

Figure 4. Measured and predicted refractive indices of 6F polyimides.

that have been observed experimentally in 6F materials can be predicted from the modified chemical structure. Furthermore, the good agreement between measured refractive index and these predictions suggests that chain packing may be reduced in these materials.

Polyimides containing pentafluorosulfanyl (SF$_5$). Recently, St. Clair and St. Clair (11) reported the synthesis of novel fluorinated polyimides containing a pentafluorosulfanyl (SF$_5$) group. Measured and predicted refractive indices of these polymers are shown in Table IV. We found a small deviation between predicted and experimental values. The least deviation occurred in a polyimide containing a bulky isopropylidene group in the dianhydride portion of the chain (IPAN). Likely, the SF$_5$ group impedes chain packing to some extent, though not as well as the 6F group.

Conclusions

Incorporation of fluorinated groups is an effective means of reducing the refractive index in polyimides. The observed reduction in refractive index likely results from a combination of reduced electronic polarizability and reduced interchain interaction and/or packing. Property prediction methods are most accurate in polyimides with bulky fluorinated substituents. Property prediction methods substantially underpredict refractive indices in non-fluorinated polyimides.

Acknowledgments

We would like to acknowledge Dr. Terry St. Clair of the NASA Langley Research Center for PMDA-ODA, ODPA-PDA, and BTDA-1,3BABB (LaRC-CPI), as well as Mr. Yasunori Sugita of the Mitsui Toatsu Chemical Company for NEW-TPI. We would also like to thank Professor Stephen Senturia of the Massachusetts Institute of Technology for the Metricon PC-2000. Finally, we thank the AT&T New Research Initiative for financial support.

Literature Cited

1. D.W. van Krevelen *Properties of Polymers*; Elsevier: Amsterdam, **1990**.
2. J. Bicerano *Prediction of Polymer Properties*; Marcel Dekker: New York, **1993**.
3. T.P. Russell; H. Gugger; J.D. Swalen *J. Polym. Sci.: Polym. Phys. Ed.* **1983** *21*, p. 1745.
4. S. Noe, PhD. Thesis, Massachusetts Institute of Technology, **1992**.
5. A.K. St. Clair; T.L. St. Clair In *Polymers for High Technology*; Eds. M.J. Bowden and S.R. Turner; American Chemical Society: Washington, D.C., **1987**; p. 428.
6. S. Herminghaus; D. Boese; D.Y. Yoon; B.A. Smith *Appl. Phys. Lett.* **1991**, *59 (9)*, p. 1043.
7. V.E. Smirnova; M.I. Bessonov In *Polyimides: Materials, Chemistry and Characterization*; Eds. C. Feger, M.M. Khojasteh, and J.E. McGrath; Elsevier Science Publishers: Amsterdam, p. 563, **1989**.
8. R. Reuter; H. Franke; C. Feger *Appl. Opt.* **1988**, *27 (21)*, p. 4565.
9. G. Hougham; G. Tesoro; A. Viehbeck; J. Chapple-Sokol *Amer. Chem. Soc. Polym. Prep.* **1993**, *34(1)*, p. 371.
10. D.C. Rich; P. Cebe; A.K. St. Clair In *MRS Symp. Proc.: Elec. Pack. Mat. Sci. VII*; Eds. P. Borgensen, K.F. Jensen, R. Pollak, and H. Yamada; Materials Research Society: Pittsburgh, PA, **1994**; *323*, p. 301.
11. A.K. St. Clair; T.L. St.Clair *Polym. Prep.* **1993**, *34(1)*, p. 385.

RECEIVED November 7, 1994

Author Index

Affiliation Index

Subject Index